Debates in Science Edu

This fully revised second edition of *Debates in Science Education* explores the major issues that science teachers encounter in teaching their subject, encouraging the reader to make their own informed judgements and argue their point of view with deeper theoretical knowledge and understanding.

Brand new chapters written by a team of international experts provide fresh insight into topics of central importance when teaching science. Written to aid and inspire beginning teachers, current teachers and established subject leaders, these focused chapters are essential to anyone wishing to deepen their understanding of salient issues within school science education, including:

- STEAM education;
- sustainability and climate change;
- science and sensitive issues;
- equity and diversity;
- science and sex education;
- science and religion;
- science and pedagogy (including science inquiry);
- transition from primary to secondary school.

Encouraging critical reflection and aiming to stimulate both novice and experienced teachers, this book is a valuable resource for any student or practicing teacher and particularly those engaged in continuing professional development or Master's level study.

Justin Dillon is Professor of Science and Environmental Education at UCL. He was President of the European Science Education Research Association from 2007–11 and is currently President of the UK National Association for Environmental Education.

Mike Watts is Professor of Education at Brunel University London. He was awarded a National Teaching Fellowship in 2005, is a Fellow of the Institute of Physics and is a council member of the National Conference of University Professors.

Debates in Subject Teaching
Series Edited by Susan Capel, Jon Davison and James Arthur

Each title in the Debates in Subject Teaching series presents high-quality material, specially commissioned to stimulate teachers engaged in initial teacher education, continuing professional development and Master's level study to think more deeply about their practice, and link research and evidence to what they have observed in schools. By providing up-to-date, comprehensive coverage the books in the series support teachers in reaching their own informed judgements, enabling them to discuss and argue their point of view with deeper theoretical knowledge and understanding.

Debates in Mathematics Education, 2nd edition
Edited by Gwen Ineson and Hilary Povey

Debates in Primary Education
Edited by Virginia Bower

Debates in Art and Design Education, 2nd edition
Edited by Nicholas Addison and Lesley Burgess

Debates in Second Language Education, 1st edition
Edited by Ernesto Macaro and Robert Woore

Debates in Science Education, 2nd edition
Edited by Justin Dillon and Mike Watts

Debates in Design and Technology Education, 2nd edition
Edited by Alison Hardy

For more information about this series, please visit: www.routledge.com/Debates-in-Subject-Teaching/book-series/DIST

Debates in Science Education

Second Edition

Edited by Justin Dillon and Mike Watts

Routledge
Taylor & Francis Group

LONDON AND NEW YORK

Cover image: © Carsten Koall / Stringer / Getty Images

Second edition published 2023
by Routledge
4 Park Square, Milton Park, Abingdon, Oxon, OX14 4RN

and by Routledge
605 Third Avenue, New York, NY 10158

Routledge is an imprint of the Taylor & Francis Group, an informa business

© 2023 selection and editorial matter, Justin Dillon and Mike Watts; individual chapters, the contributors

First edition published by Routledge 2014

British Library Cataloguing-in-Publication Data
A catalogue record for this book is available from the British Library

Library of Congress Cataloging-in-Publication Data
Names: Dillon, Justin, editor. | Watts, Mike, editor.
Title: Debates in science education / edited by Justin Dillon and Mike Watts.
Description: Second edition. | Abingdon, Oxon ; New York, NY : Routledge, 2023. | Series: Debates in subject teaching | Includes bibliographical references and index.
Identifiers: LCCN 2022014973 (print) | LCCN 2022014974 (ebook) | ISBN 9780367685157 (paperback) | ISBN 9780367685140 (hardback) | ISBN 9781003137894 (ebook)
Subjects: LCSH: Science—Study and teaching. | Science teachers—In-service training.
Classification: LCC LB1585 .D39 2023 (print) | LCC LB1585 (ebook) | DDC 507.1—dc23/eng/20220330
LC record available at https://lccn.loc.gov/2022014973
LC ebook record available at https://lccn.loc.gov/2022014974

ISBN: 978-0-367-68514-0 (hbk)
ISBN: 978-0-367-68515-7 (pbk)
ISBN: 978-1-003-13789-4 (ebk)

DOI: 10.4324/9781003137894

Typeset in Galliard
by Apex CoVantage, LLC

Contents

Contributors

Steve Alsop is Professor of Education at York University, Toronto, Canada, where he teaches courses and supervises graduate students in education, science and technology studies and environmental studies. His research explores science pedagogies in a variety of public educational contexts, including schools, universities, museums and environmental organisations. Steve's recent work is located in the Dadaab refugee camp in Kenya with the Borderless Higher Education for Refugees programme. His teaching, writing and research envisage education as a collaborative process of building more wonderous, caring, responsively-diverse, equitable worlds.

Larry Bencze is Associate Professor (Emeritus) in Science Education at the University of Toronto, Canada (1998–present). Prior to this role, he worked for 15 years as a science teacher and as a science education consultant in Ontario. His research programme emphasizes critical analyses – drawing on history, philosophy, sociology, etc. – of science and technology, explicit teaching about problematic power relations and student-led, research-informed and negotiated socio-political actions to address personal, social and environmental harms associated with fields of science and technology. Recent publications include two edited books about proactive citizenship. He also is co-editor of an open-source activist journal (goo.gl/ir7YRj).

Jenny Byrne is Associate Professor in Education at Southampton Education School, the University of Southampton, UK. Her expertise and research interests include health education and exploring the connections between science and health education, including scientific and health literacy, science education in formal and informal settings and socio-scientific issues. Jenny graduated from Birmingham University with a BSc in Bacteriology and completed a PGCE at Durham University. She taught in secondary schools and became Head of Science. Jenny has subsequently taught in all phases of education including a pupil referral unit, and as a health education officer and adviser for before moving to higher education. Since then, she has taught and led undergraduate and PGCE primary and secondary science programmes, as well as supervised Master's and PhD students.

Ann Childs is Associate Professor in Science Education at the University of Oxford, UK. Her research focuses on the professional development of science teachers and teacher educators. Ann completed her PhD in chemistry at Birmingham University in 1982 and then trained to be a science teacher at Oxford University. She taught for 11 years in Oxfordshire and in Sierra Leone for Voluntary Services Overseas (VSO). During her work as a teacher in Oxfordshire, Ann mentored beginning science teachers on the Oxford Internship Scheme. She took up her current post in 1997 where she now teaches on the PGCE, is director of the Master's in Teacher Education and supervises Master's and DPhil students.

Laura Colucci-Gray is Senior Lecturer in Science and Sustainability Education and Head of the Institute of Education, Teaching and Leadership at Moray House School of Education and Sport, University of Edinburgh, UK. Her research focuses on science-society debates and the development of participatory approaches engaging artistic and scientific creativities for radical democracy. After gaining her first degree in Natural Sciences at the University of Turin, Laura taught Biology in a secondary school in Italy. She then moved to the UK, where she gained a PhD in Science Education from the Open University (UK) and worked as a lecturer and Director of Research at the University of Aberdeen until 2017. Laura has been President of the Scottish Educational Research Association and she is a Visiting Professor in STEAM Education at the University of Turin.

Justin Dillon is Professor of Science and Environmental Education at University College London, UK. His research focuses on learning and engagement in and out of schools. After studying for a degree in chemistry, Justin trained as a teacher and taught in London schools before joining King's College London in 1989, where he worked as a researcher and teacher educator until 2014. Justin was President of the European Science Education Research Association (ESERA) from 2007–11 and is President of the UK National Association for Environmental Education. He edits the journal *Studies in Science Education* and is an editor of the *International Journal of Science Education*.

Sarah Earle is Reader in Education at Bath Spa University, UK. She was a primary school teacher for 13 years before moving into initial teacher education in 2012. Sarah's PhD considered formative and summative assessment in primary science. Since 2015, she has led the Teacher Assessment in Primary Science (TAPS) project, co-researching with teachers in each of the four nations of the UK to develop support for primary science practice. Sarah supports the professional learning of class teachers and science subject leaders via online and face-to-face courses and through her work with the Association for Science Education and the Primary Science Quality Mark. She is also editor of the open-access *Journal of Emergent Science*.

Erin Marie Furtak is Professor of STEM Education in the School of Education at the University of Colorado, Boulder, USA. After receiving an undergraduate degree in Biology, she worked as a public high school science teacher before pursuing her PhD in Curriculum and Teacher Education at Stanford University. Erin's research investigates the ways that secondary science teachers design and enact formative assessments, how this process informs teachers' learning and, in turn, how improvements in teachers' formative assessment practice over support student learning. She currently directs a long-term research-practice partnership, funded by the National Science Foundation and the Spencer Foundation, with a large and economically, culturally, linguistically and ethnically diverse school district focused on supporting high school teachers' classroom assessment practices.

Robyn M. Gillies is Professor of Education at the University of Queensland, Brisbane, Australia. Her research focuses on the social and cognitive aspects of learning through social interaction with a particular focus on the learning sciences, classroom discourses and small group processes. Robyn has worked extensively in both primary and secondary schools to embed STEM education initiatives into the science curriculum.

Lindsay Hetherington is Associate Professor of Science Education at the University of Exeter, UK. She researches socio-material approaches to science education and creativity in science education. Lindsay completed a degree in Natural Sciences and PGCE in Chemistry before working as a science teacher and Head of Chemistry. In 2006, she moved to the University of Exeter, where she has taught PGCE, MA Education, EdD and PhD students and is currently Head of Initial Teacher Education. Lindsay is co-editor of the journal *Research in Science and Technological Education.*

Rosária Justi is Professor of Science Education at the Universidade Federal de Minas Gerais (UFMG), Brazil. Her research interests are related to modelling-based science education, the nature of science and science teachers' education. Rosária has a bachelor's degree in Chemistry, a Master's in Education and a doctorate in Science Education. She has also conducted post-doctorate projects in collaboration with colleagues from Dutch and British universities. Before joining UFMG, where Rosária worked as teacher educator, she taught chemistry in secondary schools in Brazil. Rosária was an editor of the *International Journal of Science Education* from 2014 to 2020, and the editor-in-chief of the *Revista Brasileira de Pesquisa em Educação em Ciências* (*Brazilian Journal of Research in Science Education*) from 2015 to 2021.

Poliana Maia has been Associate Professor of Chemistry Education at the Universidade Federal de Viçosa, Brazil, since 2008. Her research focuses on models and modelling, the nature of science and teacher education. Poliana graduated with a degree in Chemistry and achieved Master's and doctoral degrees in Studies in Education. Her postdoctoral project was about teacher education, focusing on the development of pedagogical content knowledge. Poliana acts as a deputy leader of a science education research group, of which she has been a member since 2000. She currently supervises Science and Maths Education Master's students.

Jonathan Osborne is Kamalachari Professor in Science Education, Emeritus, in the Graduate School of Education, Stanford University, USA. He was President of the US National Association for Research in Science Teaching (2006–07) and has won the Association's award for the best research publication in the *Journal of Research in Science Teaching* twice (2003 and 2004), and the Distinguished Contribution to Science Education Award in 2018. Jonathan was a member of the US National Academies Panel that produced the *Framework for K-12 Science Education*. He has also chaired the expert group for the science assessments conducted by the OECD PISA for 2015 and for the forthcoming one in 2025.

Michael J. Reiss is Professor of Science Education at University College London, UK, Honorary Visiting Professor at the Royal Veterinary College, Honorary Fellow of the British Science Association, Fellow of the Academy of Social Sciences, a member of the Nuffield Council on Bioethics, President of the International Society for Science and Religion and President of the Association for Science Education. After a PhD and post-doc in Evolutionary Biology, he trained as a secondary teacher and taught in schools for five years. Michael then returned to higher education, spending six years on secondary teacher training and six years on primary teacher training before taking up his present post in 2001.

Saima Salehjee is Senior Lecturer in Education at the University of Glasgow, UK. Her research focuses on science literacy and identity formation among multi-aged people from different ethnic, religious and sexual backgrounds. Saima has received research funding from the Royal Society of Chemistry to continue her research with children on the provision of science literacy using stories as a 'nudge' in science teaching and learning practices. In addition, she has received funding from Skills Development Scotland to support the development of teacher-researchers in further education. Saima is an active member of the Advance HE anti-racist curriculum project, helping staff and students to feel confident and supported throughout their higher education journey.

Monique Santos is a Science Education PhD student at the Universidade Federal de Minas Gerais, Brazil. She has a Master's in Science Education and a chemistry teacher certificate from the same university. She also studied at Università degli Studi di Roma Tor Vergata in Italy as part of her teacher education. Monique has been teaching chemistry in Brazilian schools for some years. Her research interests include history, philosophy, sociology and nature of science; modelling-based science education; and teachers' knowledge.

Mike Watts is Professor of Education at Brunel University London, UK. He has a background in Physics, and taught secondary school science for many years in Hackney, north London, and Kingston, Jamaica, before undertaking a PhD at the University of Surrey. He was awarded a National Teaching Fellowship in 2005, joined Brunel University London in 2006, is a Fellow of the Institute of Physics and is a council member of the National Conference of University Professors. Mike teaches at all levels within Brunel's Department of Education, has published widely and enjoys research in the scholarship of teaching and learning in higher education, the public understanding of science and the use of narrative story-telling in qualitative education research.

Victoria Wong is a Senior Lecturer in Science Education at the University of Exeter. Her research interests include how science education policy is made and enacted and students' use of mathematics within science. After studying for a degree in Chemistry, Victoria trained as a teacher and taught science and chemistry in England, Spain and New Zealand. She has also worked as an independent science education consultant for organisations including the Royal Society of Chemistry and the Nuffield Foundation, and as a teacher educator at King's College London and the University of Oxford, UK. In 2020, Victoria returned to the classroom to gain some up-to-date experience and taught science in a state school in England before joining the University of Exeter in 2022.

Introduction to the series

This book, *Debates in Science Education*, is one of a series entitled Debates in Subject Teaching, many of which are now in their second and third editions. The series has been designed to engage with a wide range of debates related to subject teaching. Unquestionably, debates vary among the subjects, but may include, for example, issues that are related to:

- the definition, purpose and aims of the subject;
- the curriculum and content of the subject;
- subject pedagogy;
- the development of the subject and its future in the 21st century;
- the relationship between the subject and broader educational aims and objectives in society, and the philosophy and sociology of education.

The outcome of these debates might, for example, support the justification for the subject, or be addressed in the classroom through the teaching of the subject and/or impact on initial teacher education and continuing professional development (CPD) in the subject.

Likewise, debates change within subjects over time. Consequently, each book presents key debates that subject teachers should understand, reflect on and engage in at the time it was written (and subsequent editions of the book are likely to include debates about different issues, as well as revisiting some enduring debates in the subject). Chapters have been designed to highlight major questions, and to consider the evidence from research and practice in order to find possible answers. Some subject books or chapters offer at least one solution or a view of the ways forward, whereas others provide alternative views and leave readers to identify their own solution or view of the ways forward. It is anticipated that readers will want to pursue the issues raised; hence, chapters include questions for further debate and suggestions for further reading. Debates covered in the series provide the basis for discussion in university subject seminars and meetings between professionals in school departmental meetings and in the context of CPD courses. The topics are also appropriate for consideration in assignments or classroom-based research. The books have been written for all those with a professional interest in the subject, including student teachers learning to teach the subject in secondary schools; newly qualified teachers; teachers undertaking study at Master's level; teachers with a subject coordination or leadership role and those preparing for such responsibility; as well as school-based mentors, university tutors and advisers of the aforementioned groups.

Because of the range of issues covered, each subject book is an edited collection. Editors have commissioned new writing from experts on particular issues for debate, which, collectively, represent many different perspectives on a subject and the teaching of the subject. Readers should not expect a book in this series to cover all aspects of a debate, cover the entire range of debates in a subject, offer a completely unified view of the subject/teaching of the subject or deal with each debate discretely. Part of what each book in this series offers to readers is the opportunity to explore the interrelationships between positions in debates and, indeed, among the debates themselves, by identifying the overlapping concerns and competing arguments that are woven through the text. Many initiatives in subject teaching continue to originate from central government, and, as a result, teachers have decreasing control of subject content, pedagogy and assessment strategies. It is strongly felt that for teaching to remain properly a vocation and a profession, teachers must be invited to be part of a creative and critical dialogue about subject teaching, and should be encouraged to reflect, criticise, problem solve and innovate. This series is intended to provide teachers with a stimulus for democratic involvement in the development of the discourse of subject teaching.

Susan Capel, Jon Davison and James Arthur
March 2019

Chapter 1

Debates in Science Education

Justin Dillon and Mike Watts

Welcome to *Debates in Science Education*. This book is designed for anyone with an interest in school science education, but specifically student/beginning teachers preparing to teach science on secondary initial teacher education courses, induction-year, newly qualified and early-career secondary science teachers in England. It is also likely to be of interest to more experienced science teachers studying for a Master's degree and/or progressing their career into subject mentoring and/or an expert teacher or subject leader in science as part of their continuing professional development. The number and variety of audiences should indicate that developing as a teacher takes time. But it also needs support, and we hope that this volume, written by a hand-picked collection of scholars, will interest and provoke you in your quest – whatever that might be.

On purpose

Both of us started our careers as teachers when science education, as an area of research, was in its relative infancy in the UK. Back in the 1960s, the Nuffield Science project galvanised science education through 'discovery learning', 'learning by doing' and 'teaching for understanding, not learning' – just at the time when one of us (MW) was beginning professional life as a secondary physics and science teacher. The first edition of the *European Journal of Science Education* was published in 1979, the year that the other (JD) started his initial teacher education at Chelsea College. There were articles on Piaget's cognitive psychology, science and technology in the classroom, teaching thermodynamics, audio-visual material and environmental education. There was also a paper by Maurice Galton and Jim Eggleston entitled 'Some characteristics of effective science teaching', which begins by noting that "From the beginning, one of the central concerns of educational enquiry has been the attempt to answer the question 'What makes a good teacher?'" (1979, p. 75).

Galton and Eggleston report on a study that involved using an instrument called the Science Teaching Observation Schedule to tally the 'intellectual transactions' that took place during science lessons. A total of 94 classes of students aged 14 and 15 were observed being taught by different teachers. Three teaching

DOI: 10.4324/9781003137894-1

styles were identified: 'Problem-solvers', 'Informers' and 'Enquirers'. 'Informers' spent more time on teacher-centred activities than 'Problem-solvers'. 'Enquirers' spent the least amount of time on such activities, preferring instead to engage their students with more practical activities and demonstrations. Various instruments were used to measure students' recall of information, as well as their ability to manipulate data and to solve problems. The authors noted:

> There was little consistent support for the effectiveness of style II teaching [in biology, chemistry and physics] . . . there were grounds for suggesting that the teacher-directed, didactic approach of the 'informing style' was by far the least effective of all the three types.
>
> (p. 82)

Jump forward to March 1, 2021 – the Secretary of State for Education, Gavin Williamson, is addressing the Foundation for Education Development, National Education Summit:

> We know much more now about what works best: evidence-backed, traditional teacher-led lessons with children seated facing the expert at the front of the class are powerful tools for enabling a structured learning environment where everyone flourishes.

There is no nuance here nor sense of there being a debate, just rhetoric. Williamson, who was widely derided as one of the worst and most unpopular education secretaries (Whittaker, 2021), was sacked six months later, just over two years after he had been sacked as defence minister for leaking secret information. Schools, teachers and students deserve better than this, and yet education consistently ends up being used as a political football.

Over the past 40-odd years, we have witnessed a number of debates: within education generally and within science education in particular. We have seen the rise, fall, rise and fall of inquiry-based approaches in science classrooms; an obsession with subject knowledge; arguments for and against integrating the sciences; a growth in the influence of neuroscience; concerns over health and safety in science laboratories; and arguments for and against large-scale international tests such as PISA. We think that it is important that teachers keep abreast of the debates and decide for themselves where they stand – this approach, we would argue, reflects our view as teachers as professionals rather than teachers as technicians.

On argument and debate

On January 22, 2016, the Government of Japan released their *5th Science and Technology Basic Plan*. This plan proposes the idea of 'Society 5.0', a vision of a future society guided by scientific and technological innovation. Society 5.0 is intended as the next stage of human development beyond the kind of societal living we experience today – it is the next 'big step forward'. So, for example, while big data analysis, the internet of things, artificial intelligence and information strategies are already regular fixtures in life, the intention is to use these to shape our future working lives, shopping habits, leisure activities, health services, education and transport mobility – to note just a few. Society 5.0 is heralded as a

> human-centred society that, through the high degree of merging between cyberspace and physical space, will be able to balance economic advancement with the resolution of social problems by providing goods and services . . . to ensure that all citizens can lead high-quality, lives full of comfort and vitality.
> (Japan Government Cabinet Office, 2017, p. 1)

One immediate question, though: is this a good thing? At a time when the UK and many other countries are emerging from the impact of a Covid pandemic, it is not difficult to join a chorus of people who wish to re-imagine society. Arguably, what is mooted as policy in Japan may – or may not – become reality elsewhere in the world. In general, though, where one country steps, others are prone to follow. But should this happen without debate? And if that debate is to happen, exactly how and where should it take place? It is clearly not solely the domain of government (Japan's or otherwise), big business, television documentaries, experts and pundits, the 'broadsheet' press – it is a debate that belongs to all parts of society.

Policymakers, scientists, NGOs and even entrepreneurs have been accentuating the need for dialogue and debate between science and citizens in order to discuss the social, ethical and legal implications of new research findings and potentially cutting-edge technologies (Burri, 2019), and Society 5.0 need be no exception to this. It is a debate that also belongs to young people in schools, colleges, universities and more generally in life, not just under the umbrella of science education. In this respect, science education provides society with a starting point to address discourses of science information, of trust (and mistrust) with and for future citizens and consumers of scientific knowledge and understanding. And it is not just a debate about scientific information alone, but also one of culture, morals and values. While science students' classroom experiences represent only a small cross-section of their life experiences, we argue that they form an important foothold in the processes of becoming science literate and 'science confident'. And we see discussion, argumentation and debate as important ingredients in this journey.

In this opening chapter, we have a clear agenda and take a distinctive line in our view of science education: in a nutshell, the more conceptual challenge to be engineered, the better. It is, after all, a book about debates. We consider first the ways in which debate and argumentation work with people (young and old), then take a look at classroom life and draw on some of the polemics and practices described in our chapters to come. We hope to address some 'so what?' questions along the way. Needless to say, the directions we take might not uniformly be deemed appropriate or acceptable by all our readers – but that too is necessarily the subject of debate.

The word debate, of course, has a series of meanings: a formally structured affair (as organised by a debating society); a serious discussion of a subject (as in a public debate about education); or an attempt to reach an everyday decision (as in 'we debated whether or not to grab a taxi to the station'). That is, the term spans a spectrum of argumentation from the very formal to the informal; in our discussion in this chapter, we give examples of several points along that continuum. But must debaters be antagonistic and take sides? The etymology of the word derives from old Anglo-French: *de batre*, to beat, from the Latin *battuere*, with the sense of beating someone into submission. So, yes, in the traditional form, proponents and opponents take opposing sides in debating a proposed motion that is 'put before the house', at the end of which there is a vote by the audience to decide which side of the argument beat the other and thereby won the day. However, outside of this formal structure, there lies a range of possibilities: from debating with friends when chewing over the pros and cons of a particular film, to deliberating with oneself in inner dialogue ('I was debating with myself whether to go out jogging in the rain'). We also signal to each other that something is contentious when we say that it is 'open to debate' or is 'the subject of considerable debate'. The common thread throughout this is the recourse to argument and counterargument, the articulation of reason, of explanation and the use of persuasive exemplars and evidence.

The International Debate Education Society (IDEBATE) comprises a worldwide network of organisations that educates young people in debate as one means of giving voice to their ideas. The society argues that debating provides a platform for them to speak up and be heard in safe and structured environments. In the UK, there is the Great Debaters Club, with a similar mission; at a more local level, there is Sylvans Debating Society, based in London's Fleet Street and founded in 1866 by luminaries of the day, such as Charles Dickens and Samuel Johnson. In a similar vein, there are numerous online courses that encourage and provide training for debate. The unifying feature of such organisations is that debate enhances education though making learning interactive. The claim is that students who had previously been disengaged and passive learners report a surge of enthusiasm when their teachers have encouraged them to question what they were learning and debate the content of their textbooks with each other, instead of simply assimilating it in preparation for exams (Avery et al., 2013; Omelicheva et al., 2008).

On science and arguments

First, there are good arguments and bad arguments. One sense of the expression 'we had a bad argument' is that the occasion was fraught, confrontational; the people involved were left disappointed and exasperated; their interactions remained unresolved; the atmosphere was uncomfortably emotional, rancorous and disagreeable. A good argument, on the other hand, is a stirring exchange of opinions and perspectives. It happens in a manageable fashion so that, at some appropriate point, a 'knot can be tied' in the debate that allows people to agree – even if only to agree to disagree. As Joseph Joubert, the French essayist, once said, "It is better to debate a question without settling it than to settle a question without debating it." People emerge from the give-and-take of a good argument with positive feelings, food for thought and even a sense of pleasure in the whole affair. There is need to recognise not only the strengths but also the limitations of one's own position, as well as the arguments of others. In this chapter, then – as we do throughout the book – we are looking to engender good arguments. That is not necessarily to tone down or iron out passions and feelings: we want good arguments to be exciting and stimulating but handled in such a manner that allows for strong expression within the realms of decency and mutual respect. Another maxim: as Bishop Desmond Tutu (2004) has said, "Don't raise your voice, improve your argument." More of this later in the chapter.

A moment, though, to discuss a model. Hodges' model was developed in the UK during the early 1980s and is a conceptual framework that is intended to be person-centred and situation-based (Jones, 2000). To begin, the question Hodges originally posed was: who are the recipients of science and health information? Well, first and foremost, individuals of all ages, races and creed, but also groups of people, families, communities and populations. Then Hodges asked: what types of activities – tasks, duties and treatments – do science educators carry out? They must always act professionally, but frequently according to strict rules and policies; their actions are often dictated by specific ways of working within schools, colleges, universities, health systems, museum complexes, field study centres and so on. Educators do many things by routine according to precise procedures, and these can be classed as 'mechanistic' – they contrast with times when educators of all kinds give of themselves to reassure, comfort, develop rapport and engage educationally. This is what Hodges described as 'humanistic': what the public commonly think of as the caring teacher. In use, Hodges' framework prompts us to consider four major subject headings or domains of knowledge. Namely, what knowledge is needed to cater for individuals and/or groups and undertake humanistic and/or mechanistic activities? Through these questions, Hodges derived the model we depict in Figures 1.1 and 1.2.

Debate in science education can take place in each and any of these quadrants. Traditional policy debates might ask students to argue for or against a specific political position; for example, the science that lies behind alcohol limits when driving a car, or the reasons behind dietitians' mantra of 'five a day' portions of fruit or vegetables. Other modes of debate could require students to discuss a variety of perspectives regarding essential concepts and values (e.g., 'freedom' and 'responsibility' in terms, say, of cigarette smoking, Covid vaccinations, owning a pet or the culling of badgers). Similarly, while many in-class debates are carefully moderated by the teacher or lecturer, other modes are less formally regulated,

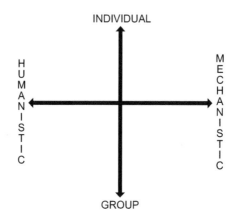

Figure 1.1 Two continuums as modes for science information (based on Jones, 2000)

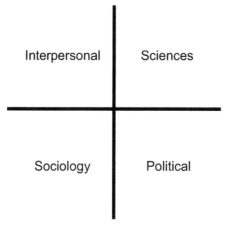

Figure 1.2 Four domains of science information (based on Jones, 2000)

and we give some examples of these in four classroom scenarios a little later. In each case, the debates can be seen to move through each of Hodges' domains, swinging between 'structured science' and interpersonal experience, between and within the of politics and social dimensions of science, between the structures and politics of science, and so on.

It is possible, then, to recognise debates that take place in academic spheres, in the national media, within educational circles, at local institutional levels and so on. These might be debates about the history or philosophy of science, of science curriculums and provision, concerns over the direct teaching of 'basic building blocks' or constructivist approaches, about pedagogies such as the virtues of 'instructional methods' over 'discovery learning' or problem-based learning, of the relationships of theory to practical work – along with a multitude more. All the coming chapters of the book aim to place such debates within areas of wide and established concern and then, where possible, to ground these in matters of educational practice. The intention in each case is to 'bring the debate home' so that even the most elevated and philosophical of issues can be seen to impact on everyday science education.

Argumentation and debate are a core part of science (Bricker & Bell, 2008). An important element of the scientific enterprise is the ability to understand and practise valid ways of arguing in a scientific context. There is a commonly held view that arguments in science are largely resolved and settled by recourse to direct evidence; being scientific implies being sceptical of any claims unsupported by clear, compelling evidence, and then rejecting or modifying claims that fit only poorly with whatever evidence is (or is not) available. From this point of view, the discipline of science is distinguished by its "central commitment to evidence as the basis of justified belief about material causes", and the rational means of resolving controversy (Siegel, 1989). This is only partly true, of course: the early quantum physicist Paul Dirac, driven entirely by the beauty of his equations, put forward his theory on the existence of anti-matter many years before evidence was derived in 1932. Erwin Schrödinger arrived at his famous equation through nothing more than inspired guesswork. Much can be said, too, of the powerful role of imaginative thought experiments that have frequently dotted the history of science. Nevertheless, evidence is vitally important, and science can generally be seen to make progress through the constant argumentative interplay between theory and experiment.

So, does debate really happen in science? Our answer is actually both yes and no. Of course, one only has to recall the historic Solvay Conferences held in Brussels in the early 1900s, which marked several epic turning points in the world of physics. In 1911, the invited-only conference participants (who included Marie Curie, Henri Poincare and the young Albert Einstein) debated *Radiation and Quanta*. In the famous fifth Solvay conference in 1927, they debated *Electrons and Photons*; the leading characters in that affair were the now-established Einstein and Niels Bohr, who debated the newly formulated quantum theory. There are many other stand-out moments in science when big issues have been debated. So, when might there be no debate in science? In his philosophical work, Imre Lakatos' (1971) theory of research programmes in the natural sciences has argued that there is a set of imperatives which determines what a research scientific programme should be, how it should unfold, how it might be defended and its scope and boundaries. Once a research programme or theory, such as quantum theory, becomes widely established, it then generates a high degree of consensus so that scientists within the programme work closely together to drive the ideas forward. It becomes a division of labour rather than a division of direction. Yes, there is some debate around the fringes, but there is often a high degree of agreement about the 'core' of the theory. In this respect, scientists do over time achieve a marked degree of accord. For anyone outside that consensus, it can be a very frustrating state of affairs. Max Planck once quipped on the very slow rate of change and acceptance of his ideas that, "A new scientific truth does not triumph by convincing its opponents and making them see the light, but rather because its opponents eventually die, and a new generation grows up that is familiar with it."

In their 2018 book, *The Enigma of Reason*, Hugo Mercier and Dan Sperber make the point that scientists use a very wide range of cognitive skills – with argumentation and debate being just two of the many. That they can achieve increasing levels of convergence on particular ideas and solutions to difficult problems derives more from the carefully constructed methods they use in shaping and evaluating scientific research than it does with the exercise of pure reason and rationality. As Mercier and Sperber say (2018, p. 173):

> Arguably science is the area of human endeavour where rationality and good reasoning are most valued. . . . Reasoning, however, does not cause scientists to spontaneously converge on the best theories. It causes them, rather, to elaborate and vigorously defend mutually incompatible competing theories. It also helps them – with a higher, even if still imperfect, degree of convergence this time – to evaluate competing theories and reach some degree of tentative agreement on the current winners of the competition.

On the future of science education

Since 2013, when the first edition of this book was published, the world order has gone through some major revisions. Awareness of the impact of human activity on climate change has grown to the extent that school students have taken part in worldwide strikes and other actions, often supported by their teachers and parents. The Covid-19 pandemic has impacted schools to the extent that online lessons became the norm for millions of students. It is becoming increasingly clear that science education has failed to deliver on its prime objective of providing a scientifically literate population. Not only have huge numbers of 'educated' people demonstrated stunning levels of ignorance about disease transmission, vaccination, weather and climate; it is obvious that people do not understand how science and scientists work.

We can carry on as normal: tinkering with the curriculum, adopting a few new pedagogical tricks to make online lessons more 'interactive', creating some new courses for science teachers, or we can realise that none of these is going to make much of a difference. We need a new vision for science education – one that realises that the environmental and health problems facing society are not going away. We need an informed debate about the role and purpose of science education – one that recognises that there are a number of issues that need to be addressed. We offer this book as part of that debate.

On content

The first edition of this book was published in 2013, so by most measures it is time for a new edition. We have not tinkered with the contents; we have totally revised them. Of the 18 authors in this edition, only two contributed to the first edition. Contributions come from Australia, Brazil, Canada, the UK and the USA. The chapters are organised into three sections: Debates about the nature and purpose of science education in contemporary society; Debates about the relationship between science and science pedagogy; and Debates about whole-school issues which have a science dimension.

Suggested further reading

Goldacre, B. (2009). *Bad science*. Fourth Estate.
Goldacre, B. (2014). *I think you'll find it's a bit more complicated than that.* Fourth Estate.
Pinker, S. (2021). *Rationality. What it is, why it seems scarce and why it matters.* Penguin Random House.

References

Avery, P. G., Levy, S. A., & Simmons, A. M. M. (2013). Deliberating controversial public issues as part of civic education. *The Social Studies, 104*(3), 105–114. https://doi.org/10.1080/00377996.2012.691571

Bricker, L., & Bell, P. (2008). Conceptualizations of argumentation from science studies and the learning sciences and their implications for the practices of science education. *Science Education, 92*, 473–498. https://doi.org/10.1002/sce.20278

Burri, T. (2019). International law and artificial intelligence. *German Yearbook of International Law, 60*(1), 91–108.

Galton, M., & Eggleston, J. (1979). Some characteristics of effective science teaching. *European Journal of Science Education, 1*(1), 75–86.

Japan Government Cabinet Office (Council for Science, Technology and Innovation). (2017). *Comprehensive strategy on science, technology and innovation (STI) for 2017.* https://www8.cao.go.jp/cstp/english/doc/2017stistrategy_main.pdf

Jones, P. (2000). *Hodges' health career care domains model, structural assumptions.* www.p-jones.demon.co.uk/theory.html

Mercier, H., & Sperber, D. (2018). *The enigma of reason.* Harvard University Press.

Omelicheya, M. L., & Avdeyeva, O. (2008). Teaching with lecture or debate? Testing the effectiveness of traditional versus active learning methods of instruction. *Political Science and Politics, 41*(3), 603–607.

Siegel, H. (1989). The rationality of science, critical thinking and science education. *Synthese, 80*(1), 9–42.

Tutu, D. (2004). *Look to the rock from which you were hewn.* Second Nelson Mandela Annual Lecture, The Nelson Mandela Theatre.

Whittaker, F. (2021). *Confirmed: Williamson sacked as education secretary in reshuffle.* https://schoolsweek.co.uk/confirmed-williamson-out-as-education-secretary-in-reshuffle/

Section 1

Debates about the
nature and purpose of
science education in
contemporary society

Chapter 2

The STEM, STEAM, STEAME debate

What does each term mean and what theoretical frameworks underpin their development?

Laura Colucci-Gray

Introduction

Over the course of the last decade, the teaching of science in schools and associated science education curricula globally, have seen the progressive incorporation of policy directives seeking to align science education with the priorities of industry and the world of work. The most common example is STEM (Science, Technology, Engineering and Mathematics), which, according to the definition of the National Science Foundation in 2001, is understood as a curricular approach integrating knowledge and skills across science-related research fields in order to drive economic competitiveness and innovation (NSF, 2020). Supplementing earlier attention on STEM, STEAM (STEM *with the addition* of the Arts) and STEAME (STEAM *with the addition* of Entrepreneurship) have more recently come to prominence in the practice of teachers – from the early years to higher education – across formal and informal contexts.

These emerging multi-disciplinary constructs may share one or more of the following features:

1   Inclusion of disciplines which may or may not be part of traditional school curricula, such as Engineering (Brophy et al., 2008).
2   Re-purposing of subjects as conventionally taught in schools by emphasising applied and economically relevant dimensions (e.g., Design and Technology Education turning into Creative Industries; Brown et al., 2011).
3   Combining academic and vocational subjects, across the sciences and the arts, through integrated approaches such as transdisciplinary creative inquiries (Colucci-Gray et al., 2019).

Across all domains, integration of disciplines across the curriculum is driven by the aspiration of a science curriculum that is both problem-focused and creativity-infused, and deemed to be more relevant to the lives of pupils (Dare et al., 2019). This chapter will discuss the extent to which such aims currently sit within science education, and make the case for a new set of perspectives and practices in science education arising through STEAM education.

DOI: 10.4324/9781003137894-3

Integrated constructs in science education

The appearance of STEM, STEAM or STEAME in school curricula globally poses significant questions about the way in which science education contributes to the production of knowledge in society. In other words, it tells us something about its curriculum. Taken from the original Latin 'currere', 'curriculum' refers to what is current and topical, but also what matters and is worthy of consideration. Intrinsic to curriculum design are questions of boundaries, which define what to include (as not everything can be included or it would not be 'current' anymore), but also which voices and which educational discourses underpin those decisions.

In the first instance, STEM-derived multidisciplinary compounds point to boundaries amongst the different disciplines – across the sciences, the arts and humanities – each one inviting their own particular ways of connecting with the world 'out there', of defining problems and setting criteria for validating knowledge claims.

In the second instance, however, STEM-derived multidisciplinary compounds also operate as discursive moves mediating top-down priorities from big players, such as governments and businesses, accounting for much of what is said and done in education (Cole, 2021). For example, in light of the growing presence of corporate-driven digital technologies in education, both STEM and STEM-derived approaches raise important questions of both content and methods in learning, which appear to bypass the choices and actions of teachers. Surfing the net and downloading videos from YouTube are not neutral actions. They are but one example of the many manifestations of the 'performative' nature of science and technological applications, with the power to affect and shape what 'counts', what is truly 'current' and what 'matters' in society (Ihde, 2012).

Following from this, debating the theoretical frameworks underpinning the emergence of such hybrid curricular constructs entails a significant interrogation of the nature of the educational process itself, specifically which qualities of knowing and being younger people should develop in current times (Burnard et al., 2021). It is widely acknowledged that we live in an age in which the idea of the 'democratic' has changed dramatically from enlarged consultation to a new ethics of engagement, one that exceeds the human in questions about who matters and whose knowledge counts. For example, changes in the ecologies of schools and universities – so evidently manifested by the Covid-19 pandemic – are provoking thinking about new ways of grasping the educational experience than those afforded by humanism (Braidotti, 2019). Positioning STEM/STEAM within the wider field of science education will thus involve grappling with very different ideas about what 'it means to know' beyond specific disciplines, and to include different communities of research and practice; each one potentially endorsing different ideas about the priorities and purposes of science education.

In this chapter, I will draw upon the concept of the 'boundary object' (Fox, 2011) as a tool for eliciting alternative discourses on science education vis a vis society and the natural world (see also Colucci-Gray et al., 2019). This preliminary review will evidence a vertical stance, largely concerned with scientific expertise driving economic growth, and a horizontal stance, concerned with enabling dialogue amongst a diversity of perspectives in extended processes of participatory inquiry. With respect to the latter, the second half of the chapter will put forward an argument for a new educational contract between science education and society, encompassing the development of critically reflexive faculties through inter- and transdisciplinary learning opportunities. Contributing a fresh perspective on recurrent debates, this stance puts forward a radical reconfiguration of current discourses in STEM and STEAM. Moving beyond intellectualist approaches, it recovers the value of felt experience as a significant dimension of a science education navigating the challenges of a complex world.

Science education across mixed purposes

Since its first inception, science education as formally taught in school has pursued competing aims. On the one hand, society's economic drivers have propelled a science education that provides the skilled workforce and technical expertise demanded by a growing economy (National Academies of Sciences, Engineering, and Medicine, 2022). On the other hand, science education has also responded to the necessity for greater participation of citizens in decision-making processes involving science and society. As a case in point, European perspectives on science education for citizenship have characteristically straddled the economic and the democratic argument (European Commission, 2015), by acknowledging the shortfall of "science-knowledgeable people" while recognising the need to increase "motivation and sense of societal responsibility to participate actively in the innovation process" (p. 7). Furthermore, science and technological applications occupy an ambiguous role in the modern imagination, both as contributing factors and sources of potential solutions to environmental issues: "As the world becomes more inter-connected and competitive and as research and technological know-how expands, new opportunities along with more complex societal challenges arise" (European Commission, 2015, p. 6).

Such debates underpin recent arguments on the need to re-think the science curriculum to extend beyond the teaching of traditional academic subjects, and beyond the boundaries of the laboratory, to support creativity and engagement of pupils with 'real' science in 'real' world settings. The process of teaching and learning would thus extend beyond transferability and acquisition of the facts of science, to the inclusion of interdisciplinarity and creativity (STEAM instead of STEM, European Commission, 2015, p. 9), but also developing new perspectives on the processes of knowing itself, for a more diverse, equitable and sustainable science education (Colucci-Gray et al., 2019).

In practical terms, the tension between these polarities has given rise to a variegated set of approaches to learning and teaching, of which we identify: (i) citizen science activities drawing on local knowledge and experiences as part of scientific research (Dillon & Lewenstein, 2011) and (ii) school-based design projects that respond to the needs of children in disadvantaged school communities (Calabrese Barton & Tan, 2009). But also, (iii) reframing the role of citizens as both 'critics and as creators' of science, such as in DIY science projects, whose practitioners are not concerned with quality or excellence, but with the production of shared realities and extended facts "by acting, tinkering and hacking in the matters that are of their care and concern" (Ferretti & Guimares-Pereira, p. 1).

All these different approaches share in the aspiration of a science education that enables pupils to interface effectively with the processes of production of scientific knowledge in society, combining specialist knowledge with an extensive range of social and practical skills. However, different educational approaches continue to belay competing positions on the nature of scientific knowledge as either product or process: how it relates with a world that is in continuous transformation; how it interfaces with value-stances and subsequently, what form it should take within the curriculum.

A view of science as doxa often prevails, whereby the world is but a representation (Eisner, 2001), and the science curriculum is instrumental to the delivery of abstract notions assessed against standards for competitive performance (Ryder, 2015). Conversely, other views of knowledge, which preceded the preoccupation with formalisms and universal abstractions of the Enlightenment period, are concerned with the particular and the qualitative, the contextual and value-based nature of knowing. This stance on knowledge underpins the views of those who believe that being scientifically literate should include both an understanding of the relationship between science and society (Cavagnetto, 2010), as well as the historical and contextual origins of the subject matter itself. Such views make room for a science education that is increasingly concerned with personal and social values, and as a result, with the realm of interpretation, culture and judgement (Biesta, 2018).

In these conditions, we will find that ideas of bringing creativity and personalisation into science education remain highly contested amongst teachers from primary and secondary sectors and across formal and informal environments (Hetherington et al., 2020). While creativity can foster learners' original thinking, increase personal engagement in the learning process, and boost motivation (Kaufman & Sternberg, 2010), a pervasive divide remains between an idea of science that explains natural observable processes in order to 'solve problems', and an idea of arts and crafts as interpretation, imagination and expression of the artist (or the pupils). If the first deals with the 'real' world, the second one is removed from it.

Science education as a boundary object

In the field of Science and Technology Studies (STS), 'boundary objects' (Star & Griesemer, 1989) are defined as constructs emerging at the intersection of multiple agencies, theoretical positions and communities of practice operating within an unequal landscape. Considering science education as a boundary object helps to make sense of the debates and dilemmas that characterise the science education community, caught between the need to preserve the canon of universal scientific knowledge on the one hand and the tension towards potentially reformist approaches, such as STEM and STEAM, on the other.

If a view of science as 'doxa' prevails, a common tension concerns the notion of 'legitimate expertise' and the extent to which localised, practical and experiential ways of knowing may be included and legitimated alongside formal, academic science (Skarlatidou et al., 2019). While this issue is commonly addressed across citizen science initiatives (Sharma et al., 2019), it reappears when claims are made about the need to design and deliver science curricula for particular types of pupils, such as 'girls' or children from ethnic minority groups, who may not be deemed sufficiently academic to engage with conventional science content. Similarly, the question of legitimisation of knowledge affects the realm of civic participation, advocacy and activism (Roche et al., 2020), propelled by experiential modes of knowing that are not explicitly recognised or elicited as part of the traditional science curriculum.

Such dividing lines of exclusion of culture and knowledge are then evidenced in STEM education policy agendas that are powerfully associated with the 'pipeline' model of education. An extractionist approach to natural resources is transferred to an extractionist view of education, tasked with the job of selecting and producing professional workers. Similarly, STEAM may 'power up' such imagery *with the addition* of the arts: by reconnecting to the steam engines of the Industrial Revolution, it also re-proposes industrial and imperialist views as the 'new' creative futures. In these particular formulations of STEM and STEAM education, there is no room for difference, and creativity is reduced to the business/mercantile model of providing a tradeable skill on the job market.

However, if boundary objects are characterised by divergent interpretations and lack of consensus, they are also sufficiently plastic to play a role in the adoption of new ideas, offering metaphors with the power to 'speak' to different communities in different ways (Fox, 2011). The task in hand is thus two-fold: (i) to clarify how far such new constructs perpetuate existing imaginaries and (ii) to tease out the extent to which they may support new sense-making practices within the science education community itself, and if so, what alternative theoretical frameworks may serve new imaginations and values.

Debating positions across STEM and STEAM

At least nine different and coexisting models appear to be used by different audiences to communicate about STEM (Bybee, 2013). In some cases, the emphasis is on generic features, such as promoting inquiry, collaboration and problem-solving skills in real-world situations. In other cases, it is the anticipated benefits that acquire importance, such as employability and competitiveness on the job market, for STEM becomes a proxy for the assumed importance of science serving the needs of the economy (Ryder, 2015). The indicators of success in STEM are thus tied to values of competitiveness and delivery of a mastery curriculum, fully aligning with the normative discourse of 'doxa' in science education. However, identifying *what* to teach exactly – and through which pedagogy – remains unknown (Bybee, 2013). As a result, STEM approaches appear devoid of content, focusing largely on the use and development of technologies (e.g., design and make a robot) in a context-based approach, with occasional applications of technology to solve real-life problems in pupils' led projects (Bati et al., 2018).

As mentioned earlier, a similar tension characterises STEAM education. Originally, STEAM policy formation in the United States was driven by a call "from industry for creative, critical thinkers, in much the same way that STEM policy sought an educational model that promised to yield a technically savvy, research driven workforce" (Allina, 2018, p. 78). However, there remained a contention as to whether raising the status of the 'A' (as a shorthand for the creative industries) to match the status of STEM subjects would necessarily translate into an improvement of science education more broadly.

Such contention manifests itself through different modalities for cross-disciplinary collaboration. For example, the arts would allegedly bring in creativity, personalisation and motivation to the teaching of science subjects (Perignat & Katz-Buonincontro, 2019), while engineering would help contextualise and integrate students' learning (Bryan et al., 2016; Breiner et al., 2012). The anticipated benefits of STEAM – based on art fusion – would thus equate to a more successful transfer of academic content, leveraging the complementarity of numerical and communicative skills (Quigley et al., 2017; Torres Gomez et al., 2021).

While this approach may appeal to the necessity of improving academic performance, and indeed significant relationships between arts engagement and academic achievement have been reported (Catterall et al., 2012; Bati et al., 2018), this view of STEAM is deeply rooted into instrumentalised views of education, perpetuating similarly instrumentalised views of technology and the arts. Their role is redefined to 'supporting creativity' as a skill confined to the service of some other attributes, be it a mathematical skill or a psychological attribute, as opposed to practices that enhance "students' creativity such as encouraging unique ideas, taking appropriate risks, learning from mistakes, and exploring new materials" (Perignat & Katz-Buonincontro, 2019, p. 32). In this way, their educational 'doings' as particular forms of knowing and learning, the experiences they may produce, as well as what they make possible for children is overshadowed by pre-defined goals and outcomes (Biesta, 2018).

"Letting the arts and science teach together"

Returning to the two lines of inquiry set out at the beginning of this chapter, referring respectively to the relationship amongst disciplines and the qualities of the educational process, alternative conceptions may branch out of a rethinking of STEAM – instead of STEM – which is grounded in the desire to promote artistic and scientific inquiry practices on equal terms. The arts will include the broader spectrum of design, computer graphics, coding, performing arts or creative problem solving – ranging from art forms to art-practices – supporting students working collaboratively on real-world applications that have no definitive solutions (Cook et al., 2020).

In illustrating a seven-step approach to designing STEAM inquiries, Park and Ko (2012) point to the development of systematic experimental skills in STEM subjects alongside seeing the 'big picture', by nurturing ethical thinking, communication, sociability, cooperation, leadership and empathy. The authors emphasise the critical application of scientific theories to different technologies in order to understand and problematise their practical impacts. For example, if we know that in physics power is defined by the amount of energy transferred per unit of time, the question of energy use can be explored both in relation to the types of technologies that are available, as well as to the political and cultural dynamics of distribution of economic power. The aim is not simply to understand how physics may be relevant to real-life applications (e.g., how industries can grow faster and bigger), but to overcome mechanical approaches to learning by engaging the philosophical and psychological nature of human beings who look to make sense of a world they themselves contribute to change (Bati et al., 2018).

Underpinning this approach to STEAM is a commitment to transdisciplinarity, a stance on knowledge that recognises system complexity: the interdependences between multiple levels of the same reality (e.g., the physical, perceptive and imaginary realms) and the inseparability of subject and object (Nicolescu, 2012). Differently from traditional disciplinary research (and education) focused on acquiring specialised knowledge within the confines of one specific system, complex thinking in transdisciplinary inquiries takes into account the interrelations with other systems and the influences of the external environment.

Examples of transdisciplinarity are more common in science than anticipated when considering, for example, the role of metaphors (Burnard et al., 2021). Concepts such as cells, power, vessels or webs not only connect different disciplines (e.g., cell is commonly used in physics, biology, computing), but they also help reveal alternative disciplinary conceptions (e.g., the cell can be seen both as a structural and functional unit). Metaphors embed the links between scientific culture and everyday practices, with the power to affect future imaginations. Transdisciplinarity does not require in-depth knowledge of all disciplines, but its focus lies on understanding how knowledge is being produced, paying enough attention to what knowledge (or product) is being made, and how and why boundaries are drawn by the different subjects with their particular methods and perspectives (Eisner, 1991).

Central to transdisciplinary inquiries therefore is dialogue, but not in the simplest sense of having a conversation, or in the more instrumental sense of cumulating facts to produce a given picture. Recent theorising around learning in transdisciplinary inquiries advocates a shift from social constructivism to radical constructivist and enactivist approaches, traced back to John Dewey's transactional theory of organism and environment (Johnson, 2007).

Core to this thinking is the notion that knowledge is not the concern of a mind or body that are independent from context, but that knowledge concerns the relationship between the activities of the organism and the consequences these activities bring about. The 'environment' in which the organism operates is thus narrowly defined by the particular ways in which every organism is physiologically and culturally set to be coordinated *with* (Vanderstraeten, 2002). For example, the hard shell of a mollusc is a protection from predators, but it is also the result of the organism's ability to coordinate its actions in the turbulence of the marine environment. The organism co-constructs itself by drawing on the affordances of that environment, by changing its own internal environment (e.g., the pumping in of minerals through the cell membranes for the construction of the shell), but in that process, the localised external environment also changes. This implies that the world is never independent of the activities of the organism, and that knowledge is always engaged in action.

Johnson (2007) refers to concepts and thoughts as "patterns of experiential interactions", as basic and recurrent structures which emerge from the sensorimotor experience of the organism encountering the world. For example, image-schemata, such as 'close' and 'warm', derive from early experiences of physical contact with a caregiver, projected into language and metaphorical thinking. Action and movement *in* the environment therefore define what enters the field of perception, what the organism pays attention to, and thus, what the organism knows. In educational terms, this means that it is not just knowledge – disciplinary or multidisciplinary – but it is the question of attention that is central to education. If the aim of science education is that of preparing for life in a changing world, then the question of pedagogy becomes *what kind* of attention. Applied to the focus of this chapter, we need to ask what kind of attention is stimulated and encouraged through STEAM approaches, as this will change our understanding of how science and art both (Patrizio, 2020) contribute to our understanding of the world with which we seek to correspond (Colucci-Gray, 2020).

What are the implications for practice?

Drawing on embodied conceptions of knowing, the dialogue between the arts and sciences can be one of alternating ways of *paying attention* (Burnard et al., 2021); that is, of exercising one's senses and perceptual abilities by modulating our activity in the environment. Through the teaching of both art history and natural history (Patrizio, 2020), the educational process will aim to teach the novice to 'see', by accepting, exploring and embracing varieties of forms of knowing in non-hierarchical ways, and bringing them into action.

One dimension of this type of learning may involve familiarisation and de-familiarisation, as students understand their 'ways of seeing' by using different tools (e.g., lenses vs. microscopes; Lindsay, 2021), and drawing organisms to acquire precise knowledge of body parts and functions (Fan, 2015; Evagorou et al., 2015; Bartoszeck & Tunnicliffe, 2017). Crucial in this process, it is the *mark making* that accompanies the modulation of seeing and provides feedback on the 'what' and 'how' of subjective perception. Lindsay (2021) illustrates learning though observational drawing in biology as the three stages of finding analogies (e.g., broom, freckles, plates), expanded analogy (e.g., using higher magnification, the 'broom' appearance is actually given by small feet with bristles), followed by formulation and testing of hypotheses (e.g., how the organism moves in different environments).

But another aspect of the learning process could also involve thinking through 'as if' and engaging imaginative modalities to allow what is being perceived to be transformed. One example may be that of projecting and imagining what 'it might be like'/'what can be known' from the perspective of organisms that are different from oneself. More than simply empathising, this imaginative process is one that develops awareness of one's thinking, of how boundaries of place and time are set. It is also a mode of thinking that can be applied to understanding processes of change, as when we are confronted with a changing landscape (e.g., after a storm or a major flood) that brings new and different 'forms' into being, and of which we may have no knowledge, familiarity or understanding (Colucci-Gray, 2020).

Thinking through 'as if' is also documented through model-making, as in experiences with children in the early years, using dance, music, theatre and materials, as "modifications of the physical" (Patrizio, 2020, p. 41). This is important educationally as it brings into awareness human preoccupation with the modification of physical substances, and in turn, how such modification affects the process of thinking. Let's take the example of making 'sand' playdough. Materials work as a reminder of the fine granularity of forms that are exposed to the milling effects of wind, tides, and turbulence, but they also bring the memory of the tickly sensation of sand between the toes. 'Being granular' emerges as a 'pattern of connection' between oneself and the environment as described by Johnson (2007).

What comes into the realm of perception therefore is not simply 'information' or 'data' as impressions made on an inert substance or a passive body (e.g., on our skin; or on the shells of other organisms). Rather, the arts as widely conceived can engender a multiplicity of different ways of seeing, hearing, feeling, and moving (Greene, 2001), underpinning a plurality of modes of listening to the world of which we are a part. A convergent mode of thinking, which harnesses the power of the arts towards the refinement of observation, formulation of hypothesis and awareness of language, can complement a divergent mode of thinking, encouraging learners to step in and out of temporary perceptions to acquire metacognition.

A teacher may be preoccupied with the type of attention, and the type of experience that modification provides, and the integrity of that process. Moving away from ideas of performance on standards, what can be valued as integrity may be the capacity for self-organisation, learning from one's own 'mistakes', or making decisions as to what to explore next (Ferretti & Guimarães Pereira, 2021).

Quoting Dewey (1934/2009, p. 22): "Experience is the result, the sign and the reward of that interaction of organism and environment which, when it is carried to the full, is a transformation of interaction into participation and communication." In this sense, also the discourse of creativity shifts from the monological – centred on the person that makes something – to the dialogical of encountering a multiplicity of different experiences across different times and spaces.

Therefore, understanding the synergy of arts and sciences, as underpinned by a radical constructivist approach to learning, emphasises the role of the body and the senses as central to knowing. However, sensorial perception is not simply a gateway of data and information put to the services of a higher brain. Good thinking, as Dewey first thought, will instead be the result of an acute sensitivity to the consequences that particular experiences of the body bring about. For example, teachers and pupils together can redesign the school environment itself to create learning spaces that support synergy of disciplines, such as physics and dance, modelling dynamic processes that support metacognition (Bassachs et al., 2020). Similarly, STEAM may be an approach that can support pupils and teachers with challenging the politics of time and space, redirecting attention to the living curriculum of growing edible plants, as in the example of STEAM gardens (Gray & Colucci-Gray, 2021). In so doing, the science education curriculum can support children's creativity and entrepreneurship, as advocated by the proponents of STEAME (Kovatcheva & Koleva, 2021) in a different key; by developing attention to one's knowing and being, it is possible to be attentive to the narrowing and expanding of boundaries, transcending the dichotomy of economic vs. ecological and democratic arguments.

Conclusions

Multiple understandings of STEM, STEAM and STEAME education exist. Tensions arise amongst competing educational discourses, each one thinking differently about what it means to reform science education and for which purpose. The focus on economic drivers – of the sciences as well as creative industries (National Science Foundation (NSF), 2020) – is resisted by those advocating "the radical democratic character" of the arts, and their struggle against plain functionality (Baldacchino, 2013, p. 354). Similarly, a discourse based on 'capital' – being that of the sciences or the arts – inevitably precludes understanding of the diversity and difference of children's and communities' own funds of knowledge (Calabrese Barton & Tan, 2009).

This chapter has made the argument that a renovated conception of STEAM education may actually reconfigure current provision in science education, by including expanded conceptions of knowing connected to experience and action in one's environment. However, what clearly emerges is that this type of education does not conform to specific models of teaching (for example, as active inquiry, discovery learning or project work). While such models are already featured in the educational world, what is suggested instead is the need to unlock new opportunities for researchers and practitioners to research into *modalities of attention*, as they are afforded in each learning environment, and how such affordances can be *modulated* through materials and exchanges with the experiences of others. Eisner (2001) referred to this attention as the development of the artistry of teaching, one that "requires sensibility, imagination, technique, and the ability to make judgments about the feel and significance of the particular" (p. 382). STEAM agendas, so defined, can thus be powerful tools in the hands of teachers who are able to *craft* experience by being attentive and perceptive to the shaping of the environment they share with the students, and which will – in turn – shape how they will relate and interact with one another, as humans and more than humans. From this point onwards, there is potential for transdisciplinary inquiries to open out new research agendas, exploring educational approaches for addressing broader issues of sustainability transitions affecting communities globally (Fortun, 2021). Letting the arts and sciences teach together can offer students and teachers the chance to be creative in the exploration of alternative ways of doing things and organising society, practicing with complex causation and creative approaches for deliberation and cooperative action. I argue that reframing STEAM through different understandings of experience can bring a radical shift in science education itself, to enable each individual child to become an expert in the process of 'seeing' and 'knowing' the other, partaking in the ongoing inquiry of how we can live together in the world.

Questions for further debate

1   What might persuade policy-makers that STEM is limiting rather than enhancing education?
2   What can be done in schools to move towards greater interdisciplinarity?
3   How can we harness STEAM approaches to go beyond dualistic understandings of cognition and emotion, science and arts, humans and the environment?

Suggested further reading

Banks, F., & Barlex, D. (2020). *Teaching STEM in the secondary school, helping teachers meet the challenge* (2nd ed.). Routledge.
Daniels, H., & Quigley, C. (2017). Exploring teachers' perceptions of STEAM through professional development: implications for science educators. *Professional Development in Education, 43*(3), 416–438.

References

Allina, B. (2018). The development of STEAM educational policy to promote student creativity and social empowerment. *Arts Education Policy Review, 119*(2), 77–87.
Baldacchino, J. (2013). What creative industries? Instrumentalism, autonomy and the education of artists. *International Journal of Education through Art, 9*(3), 343–356.

Bartoszeck, A. B., & Tunnicliffe, S. D. (2017). Development of biological literacy through drawing organisms. In P. Katz (Ed.), *Drawing for science education: An international perspective* (pp. 55–65). Sense Publishers.

Bassachs, M., Cañabate, D., Nogué, L., Serra, T., Bubnys, R., & Colomer, J. (2020). Fostering critical reflection in primary education through STEAM approaches. *Education Sciences, 10*, 384.

Bati, K., Yetişir, M. I., Çalişkan, I., Güneş, G., Saçan, E. G., & Chapman, D. L. (Reviewing Ed.). (2018). Teaching the concept of time: A steam-based program on computational thinking in science education. *Cogent Education, 5*(1), 1507306.

Biesta, G. (2018). What if? Art education beyond expression and creativity. In C. Naughton, G. Biesta, & D. R. Cole (Eds.), *Art, artists and pedagogy* (pp. 11–21). Routledge.

Braidotti, R. (2019). A theoretical framework for the critical posthumanities. *Theory, Culture & Society, 36*(6), 31–61.

Breiner, J. M., Harkness, S. S., Johnson, C. C., & Koehler, C. M. (2012). What is STEM? A discussion about conceptions of STEM in education and partnerships. *School Science and Mathematics, 112*(1), 3–11.

Brophy, S., Klein, S., Portsmore, M., & Rogers, C. (2008). Advancing engineering education in P-12 classrooms. *Journal of Engineering Education, 97*(3), 369–387.

Brown, R., Brown, J., Reardon, K., & Merrill, C. (2011). Understanding STEM: Current perceptions. *Technology and Engineering Teacher, 20*(6), 5–9.

Bryan, L. A., Moore, T. J., Johnson, C. C., & Roehrig, G. H. (2016). Integrated STEM education. In C. C. Johnson, T. J. Moore, & E. E. Peters-Burton (Eds.), *STEM roadmap: A framework for integrated STEM education* (pp. 23–37). Routledge.

Burnard, P., Colucci-Gray, L., & Sinha, P. (2021). Transdisciplinarity: Letting arts and science teach together. *Curriculum Perspectives, 41*, 113–118.

Bybee, R. W. (2013). *A case for STEM education*. NSTA Press.

Calabrese Barton, A., & Tan, E. (2009). Funds of knowledge and discourses and hybrid space. *Journal of Research in Science Teaching, 46*(1), 50–73.

Catterall, J. S., Dumais, S. A., & Hampden-Thompson, G. (2012). The arts and achievement in at-risk youth: Findings from four longitudinal studies (No. Research Report #55). National Endowment for the Arts.

Cavagnetto, A. R. (2010). Argument to foster scientific literacy: A review of argument interventions in K-12 science contexts. *Millennium, 80*(3), 431–434.

Cole, D. (2021). A new science of contemporary educational theory, practice and research. In K. Murris (Ed.), *Navigating the post-qualitative, new materialist and critical posthumanist terrain across discipline* (pp. 99–117). Routledge.

Colucci-Gray, L. (2020). Developing an ecological view through STEAM pedagogies in science education. In P. Burnard & L. Colucci-Gray (Eds.), *Why science and art creativities matter. (Re-)configuring STEAM for future-making education* (pp. 105–131). Brill Publisher.

Colucci-Gray, L., Burnard, P., Gray, D., & Cooke, C. (2019). *A critical review of STEAM (Science, Technology, Engineering, Arts, and Mathematics)*. Oxford Research Encyclopedia of Education.

Cook, K., Bush, S., Cox, R., & Edelen, D. (2020). Development of elementary teachers' science, technology, engineering, arts, and mathematics planning practices. *School Science and Mathematics, 120*(4), 197–208.

Dare, E., Ring-Whalen, E., & Roehrig, G. H. (2019). Creating a continuum of STEM models: Exploring how K-12 science teachers conceptualize STEM education. *International Journal of Science Education, 41*(12), 1701–1720.

Dewey, J. (2009). *Art as experience*. Perigee Books. (Original work published 1934)

Dillon, J., & Lewenstein, B. (2011). Citizen science: Innovation in environmental education research: Emerging trans-disciplinary perspectives. In B. Lewenstein (Ed.), *Innovation in environmental education research: Emerging trans-disciplinary perspectives*. Peter Lang International Academic Publishers.

Eisner, E. (1991). What really counts in schools. *Educational Leadership, 48*(5), 10–11, 14–17.

Eisner, E. (2001). From episteme to phronesis to artistry in the study and improvement of teaching. *Teaching and Teacher Education, 18*, 375–385.

European Commission. (2015). *Science education for responsible citizenship*. Report to the European Commission of the expert group on science education. Directorate-General for Research and Innovation. Science With and For Society. EUR 26893 EN.

Evagorou, M., Erduran, S., & Mäntylä, T. (2015). The role of visual representations in scientific practices: From conceptual understanding and knowledge generation to 'seeing' how science works. *International Journal of STEM Education, 2*(1), 1–14.

Fan, J. E. (2015). Drawing to learn: How producing graphical representations enhances scientific thinking. *Translational Issues in Psychological Science, 1*(2), 170–181.

Ferretti, F., & Guimarães Pereira, A. (2021). A new ethos for science? Exploring emerging DIY science "qualities". *Futures, 125.* https://doi.org/10.1016/j.futures.2020.102653

Fortun, K. (2021, May). "Teaching environmental teachers", contribution to GTI Forum "The pedagogy of transition". *Great Transition Initiative.* https://greattransition.org/gti-forum/pedagogy-transition-fortun

Fox, N. (2011). Boundary objects, social meanings and the success of new technologies. *Sociology, 45*(1), 70–85.

Gray, D., & Colucci-Gray, L. (2021). Cultivating primary creativities in STEAM gardens. In P. Burnard & M. Loughrey (Eds.), *Sculpting new creativities in primary education*. Routledge.

Greene, M. (2001). *Variations on a blue guitar: The Lincoln Center Institute lectures on aesthetic education*. Teachers College Press.

Hetherington, L., Chappell, K., Ruck Keene, H., Wren, H., Cukurova, M., Hathaway, C., Sotiriou, S., & Bogner, F. (2020). International educators' perspectives on the purpose of science education and the relationship between school science and creativity. *Research in Science & Technological Education, 38*(1), 19–41.

Ihde, D. (2012). Can continental philosophy deal with the new technologies? *Journal of Speculative Philosophy, 26*(2), 321–332.

Johnson, M. (2007). *The meaning of the body. Aesthetics of human understanding*. The University of Chicago Press.

Kaufman, J. C., & Sternberg, R. J. (2010). *The Cambridge handbook of creativity*. Cambridge University Press.

Kovatcheva, M., & Koleva, M. (2021). STEAME model in action: Challenges and solutions in mastering the digital culture. In M. Mahruf & C. Shohel (Eds.),

E-learning and digital education in the twenty-first century – challenges and prospects. InTech Open.

Lindsay, S. M. (2021). Integrating microscopy, art, and humanities to power STEAM learning in biology. *Invertebrate Biology, 140*(1), 1–12. https://doi.org/10.1111/ivb.12327

National Academies of Sciences, Engineering, and Medicine. (2022). *Science and engineering in preschool through elementary grades: The brilliance of children and the strengths of educators.* The National Academies Press. https://doi.org/10.17226/26215

National Science Foundation. (2020, May). *STEM education for the future. A visioning report.* NSF.

Nicolescu, B. (Ed.). (2012). *Transdisciplinarity and sustainability.* The ATLAS Publishing.

Park, N., & Ko, Y. (2012). Computer education's teaching-learning methods using educational programming language based on STEAM education. In J. J. Park, A. Zomaya, S.-S. Yeo, & S. Sahni (Eds.), *9th International conference on network and parallel computing (NPC). Sep 2012 lecture notes in computer science. LNCS-7513. Network and parallel computing* (pp. 320–327). Springer.

Patrizio, A. (2020). *The ecological eye. Assembling and eco-critical art history.* Manchester University Press.

Perignat, E., & Katz-Buonincontro, J. (2019). STEAM in practice and research: An integrative literature review. *Thinking Skills and Creativity, 31*, 31–43.

Quigley, C. F., Herro, D., & Jamil, F. M. (2017). Developing a conceptual model of STEAM teaching practices. *School Science and Mathematics, 117*(1–2), 1–12.

Roche, J., Bell, L., Galvão, C., Golumbic, Y. N., Kloetzer, L., Knoben, N., Laakso, M., Lorke, J., Mannion, G., Massetti, L., Mauchline, A., Pata, K., Ruck, A., Taraba, P., & Winter, S. (2020). Citizen science, education, and learning: Challenges and opportunities. *Frontiers in Sociology, 5.*

Ryder, J. (2015). Being professional: Accountability and authority in teachers' responses to science curriculum reform. *Studies in Science Education, 51*(1), 87–120.

Sharma, N., Colucci-Gray, L., Siddharthan, A., Comont, R., & Van der Wal, R. (2019). Designing online species identification tools for biological recording: The impact on data quality and citizen science learning. *PeerJ, 6*, e5965. https://doi.org/10.7717/peerj.5965

Skarlatidou, A., Ponti, M., Sprinks, J., Nold, C., Haklay, M., & Kanjo, E. (2019). User experience of digital technologies in citizen science. *JCOM, 18*(1). https://doi.org/10.22323/2.18010501

Star, S., & Griesemer, J. (1989). Institutional ecology, 'translations' and boundary objects: Amateurs and professionals in Berkeley's Museum of Vertebrate Zoology 1907–1939. *Social Studies of Science, 19*(3), 387–420.

Torres Gomez, J., Rodriguez-Hidalgo, A., Jerez Naranjo, Y. V., & Pelaez-Moreno, C. (2021). Teaching differently: The digital signal processing of multimedia content through the use of liberal arts. *IEEE Signal Processing Magazine, 38*(3), 94–104. https://doi.org/10.1109/MSP.2021.3053218

Vanderstraeten, R. (2002). Dewey's transactional constructivism. *Journal of Philosophy of Education, 36*(2), 233–246.

Chapter 3

To be debated

Teachers should mobilize science students to help replace capitalism

Larry Bencze

Introduction

It seems clear that many scientists, engineers and other related professionals often must orient their work towards capitalist values. Many of these values, while generating many useful products and services, prioritize private profit
over wellbeing of many individuals, societies and/or environments. This, perhaps obviously, could lead to questions like, 'To what extent should teachers of
science and engineering (and related fields) reveal such apparently-problematic
pro-capitalist orientations?,' 'Should such educators portray professional science
and engineering as highly logical and systematic and unproblematic regarding
possible harms to individuals, societies and environments; that is, representations
that may encourage more students to pursue further education in these fields?'
and, 'If professional science and engineering are to be portrayed as – to varying
extents – prioritizing capitalist over most community members' values, what may
be effective ways to do so?' In this chapter, after a relatively brief review of roles
for values in science and engineering (and related fields), suggestions are critically
discussed, with examples, of approaches for encouraging and enabling school science and engineering (or technology, etc.) to effectively negotiate roles of values
in these professional fields.

Values in science and engineering

Science educators and others appear to vary in their positions regarding roles of
values in the sciences and related fields. A convenient – although not necessarily
fully accurate – way to envisage such different views about science is Loving's
(1991) *Scientific Theory Profile* (STP). Based on her analyses of claims of different philosophers, scientists, etc., about the nature of knowledge generation (and
dissemination) in fields of science, she categorized their claims along two *epistemological* (about knowledge building) spectra, summarized in Figure 3.1. Her
analyses also were, however, largely *axiological* (Creswell & Poth, 2018); that is,
studies of beliefs that different experts held about roles of ethical, aesthetic and
other *values* that scientists commonly use in knowledge generation and truth

DOI: 10.4324/9781003137894-4

claim decisions. Extreme *Rationalists*, for instance, were said to believe that scientists are capable of avoiding non-logical and non-systematic influences on their decisions – such as values about gender, race, culture, economics, politics, etc. – and that extreme *Naturalists*, by contrast, may claim are *unavoidable* influences in the sciences. Such differences may not, however, be entirely distinct. Wittgenstein (1958), arguably one of the most prominent philosophers of the 20th century, said that *all* thoughts and actions are *value-laden* – including those of all scientists, engineers, etc., and those who study them. So, for example, believing that science methods should be unemotional is a value position. Having acknowledged that caveat, it nevertheless appears that different positions within the STP may be associated with important contestable value systems.

There is much argument and evidence to suggest that many scientists, science educators, media professionals and others support – more or less – *Rationalist-Realist* views (Figure 3.1) about science. Hodson (1998) has suggested, for example, that there is much support for norms (or *institutional imperatives*) promoted by Robert Merton (1973). Responding to cases like abuse of sciences by military, politicians and others in the Nazi regime, and Galileo's much earlier

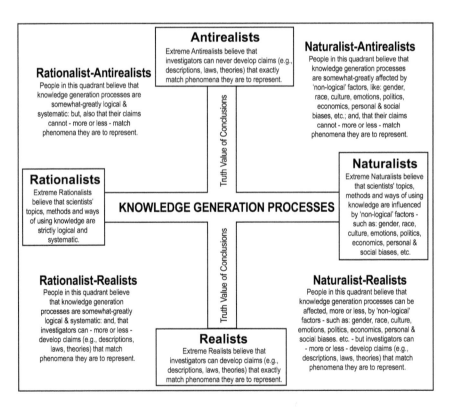

Figure 3.1 Spectra of nature of science conceptions

(c. 1610) censure by the Roman Catholic Church for promoting a heliocentric conception of our solar system (Losee, 2001), he argued that, to maintain independence from 'outsiders' and their integrity as expert knowledge producers, they needed to adhere to (and promote) a 'scientific *ethos*' that included:

- *Communalism*: As a collective activity, scientists should share ideas, methods, findings, etc., with other scientists in their communities;
- *Universalism*: All scientists, regardless of such differences as age, career status, gender, race and cultural background, can participate equally in science – including through generation and publication of findings;
- *Disinterestedness*: All scientists, along with those who support them, are expected to operate in unbiased, objective ways;
- *Originality*: Scientists are expected to contribute novel ideas, methods and results to the literature; not merely copy the work of others; and
- *Skepticism*: The scientific community is continually expected to critically scrutinize their own and colleagues' work with respect to its scientific merit, level of universalism, disinterestedness and originality.

(Ziman, 2000, pp. 57–82)

Conveniently, the acronym of these norms is CUDOS, which may suggest that those engaging in such *Rationalist* knowledge generation practices are likely to receive *kudos*, and with such visions of grandeur, be enticed to continuously support them. Related to these, it is apparent that scientists' faith in such methods to achieve truths (more *Realist* epistemological values) can greatly serve as incentives for knowledge generation pursuits in the sciences (Hodson, 1998). With widespread acceptance of such images of practices in and achievements from the sciences, professionals in these fields could have pride in their professional standards of practice, which could, in turn, justify their functioning free of outside influences.

Although there may be much acceptance of more Rationalist-Realist views about science that can justify claims for their independence from outside pressures, the 'realities' of practice appear much more complex and moreover, possibly problematic. In their two-year study of research and publication work in a prestigious research laboratory (Salk Institute for Biological Studies), Latour and Woolgar (1979/1986) suggested that investigators tended to use personal (often theory-based) biases to defend their favoured claims about the world. Also, they suggested that defences of their favoured claims involved alliances among myriad living and nonliving entities ('actants'), including: reputations of investigators, attractive structures and arrangements of graphs, 'elegance' of theories, statuses of instrumentation, etc. In their studies of *inscriptions*, such as tables and graphs, for instance, they noted that their *semiotic* features – including through strategic choices for scales of graphs' axes – could help to *conscript* other scientists into supporting their claims (McGinn & Roth, 1999).

Overall, Latour and Woolgar (1979/1986) suggested that claims from scientists are not so much data-based and rational as *socially*-constructed. Having said that, Latour (2005), with others, was instrumental in suggesting that decisions are *material-semiotic*. That is, in terms of *actor-network theory*, in the development of which he had a key part, he suggested that all decisions – including those in the sciences – are results of complex, dynamic, relationships among living (e.g., scientists), nonliving (e.g., instrumentation) and symbolic (e.g., 'elegance' of a model) entities ('actants'). This theory may explain, for instance, why some possible explanations of phenomena – such as plate tectonic theory – struggled (e.g., for decades) to gain acceptance across relevant communities of scientists, largely because of belief systems embedded in the dominant *paradigm* (Kuhn, 1970) that consisted of resilient assemblages of value-laden actants like committed scientists, predictive theories, instruments, investigation protocols, writing styles, etc.

Networks of actants may be more or less held together by many different sets of values. Foucault (2008) referred to such sets of generally cooperating actants as *dispositifs*. Among ideological perspectives that may influence assemblage of actants into dispositifs, it is apparent that few rival those associated with *capitalism* – which tends to prioritize values like: individual competitiveness; personal possessiveness; meritocracy (e.g., prestige with financial success); continuous growth; and cost externalization (arranging for others to pay costs, such as for labour) (McMurtry, 1999). Indeed, particularly since about 1970, when *neoliberal* capitalism was gaining traction, it is apparent that most living and nonliving entities on Earth have largely been 'assimilated' into a pro-capitalist *dispositif* that seems like *Star Trek*'s 'The Borg'. In contrast to earlier capitalist forms, neoliberalism tends to *favour* interventions from governments, supranational groups (e.g., World Trade Organization, International Monetary Fund, media organizations and think tanks like the Atlas Network) to help distribute pro-capitalist values among myriad actants (Cahill et al., 2018). Especially in democracies that tend not to prioritize uses of police and military to control populations, a major aspect of capitalist influence and resilience is *governmentality*; that is, conditions in which people believe they are *self*-governing, but are largely enacting values of powerful others (Foucault, 2008). Such control is apparently particularly powerful when its values have been *normalized*; that is, become subconscious assumptions about 'normal' thoughts and behaviours. Educators adhering to Rationalist-Realist views about science may, for instance, unconsciously assume that student decisions about explanations of phenomena must be supported by systematically derived experimental evidence.

Among entities assimilated into neoliberal *dispositifs*, fields of science and technology are particularly important. As depicted in Figure 3.2, development of representations (e.g., graphs) of phenomena of the world is largely attributed to the sciences. Fields of 'technology' (and engineering), on the other hand, are thought to develop 'products' (e.g., inventions, innovations, etc.) that can change the world and, moreover, generate profit (Bencze, 2020). Assuming mathematics is involved in many processes in World ←→ Sign relationships, the schema in Figure 3.2 may represent 'STEM' (science, technology, engineering

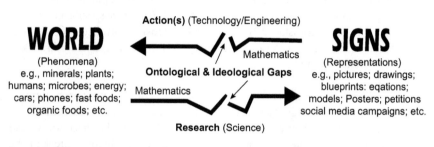

Figure 3.2 A model of STEM fields

and mathematics) fields – relationships that have become highly promoted in many educational contexts in the last two to three decades (Li et al., 2020).

There is much evidence-based argumentation suggesting apparent needs for pro-capitalist financial support for science and technology – in both academic and commercial contexts – for example, because of expenses for equipment, software, material, transportation, communication and more (Dzisah, 2007). Several analysts, however, have cautioned that for-profit arrangements among financiers, corporations, science and technology professionals, university administrations and many more entities often have compromised topic choices, investigative and analysis methods, reporting and products' uses (e.g., Grant, 2018; Krimsky, 2019; Mirowski, 2011; Ziman, 2000). There are suggestions, for instance, that pro-capitalist influences on research and development choices in some fields of science and technology have contributed to emergence of harmful viruses, like those associated with the Covid-19 pandemic, and societies' (perhaps lack of) preparedness for dealing with such socio-biological crises (Davis, 2020). More broadly, although there are strong suggestions that investigators do not always follow more Rationalist methods of science, such as hypothesizing prior to investigating (Feyerabend, 1975), it appears that many pro-capitalist science and technology activities prioritize values that seem antithetical to Merton's norms. Ziman (2000), for instance, suggested that the ethos of academic science has evolved to become more:

- *Proprietary:* i.e., not necessarily made public;
- *Local:* i.e., focused on local technical problems, rather than on general understanding;
- *Authoritarian:* i.e., often governed by outsiders, such a business personnel, rather than by individual scientists and their contextual results, etc.;
- *Commissioned:* in the sense that outsiders generally pre-specify goals to be achieved; and
- *Expert:* i.e., scientists are valued for their expert problem-solving skills, rather than for their creativity.

(pp. 78–79)

In other words, rather than valuing CUDOS/kudos, pro-capitalist science and technology fields often appear to emphasize PLACE (or, more specifically, private profit). In this vein, Krimsky (2019), who has extensively studied pro-capitalist fields of science and technology (also see: https://sites.tufts.edu/sheldonkrimsky/corrupted-science), stated:

> One cannot make a blanket judgement that corporate external funding always biases research. However, in certain fields where the commercial stakes are high, such as agricultural chemicals, tobacco, pharmaceuticals, and climate change studies, we have seen examples of the biasing effect of the sponsoring entity, ghostwriting [manipulated research articles authored by reputable scientists], withholding public health data, and fabricated attacks on responsible scientists.
>
> (p. xxxix)

Ecojust science and engineering education

Fields of science and technology have been associated with many positive developments, such as in terms of prolonging lives through advances in medical and agricultural fields. Having said that, it also appears that many such fields are associated with numerous harms to wellbeing of many individuals, societies and environments (WISE). Indeed, Ord (2020) has warned that humanity is on a virtual 'precipice', with existential threats like those linked to climate change, nuclear warfare and industrial degradation of habitats associated with unprecedented species declines and losses. We also are plagued by ongoing harms from manufactured foods, poorly tested medications and increasing electronic surveillance systems. Although all such harms can, in a sense, be blamed on large-scale *dispositifs*, many analysts strongly associate them with *neoliberal* capitalism – largely characterized by government and transnational organizations' facilitation of legal, technological, discourse-based and other conditions favouring wealth concentration into hands of few financiers and corporations, largely at expense of WISE of most other entities (e.g., Grant, 2018; Klein, 2014; Krimsky, 2019; McMurtry, 1999; Piketty, 2020).

In light of personal, social and environmental harms linked to – although *not* necessarily fully attributable to – fields of science and technology, a key debate teachers may have in their own minds and/or with colleagues, administrators and others could pertain to extents to which approaches should be used to educate students about harms linked to neoliberal science and technology, and to prepare them to develop and implement plans of action to overcome harms of their choice. Simplistically, this may be considered in terms of two opposing depictions of professional science and technology as represented in Loving's (1991) *Scientific Theory Profile* (Figure 3.1); that is, regarding choices between more *Rationalist-Realist* (RR) vs. more *Naturalist-Antirealist* (NA) depictions.

Although associations are not necessarily 'tight' (e.g., certain), teachers supporting more RR perspectives may emphasize – with reference to Lock's (1990) learning control model (Figure 3.3) – more teacher-directed (TD) and closed-ended (CE) approaches. By contrast, teachers supporting more NA views about science may accommodate more student-directed (SD) and open-ended (OE) activities, such as student-controlled research (Bencze et al., 2006). Supporters of *Rationalist-Realist* views may, for instance, believe that there are widely agreed-upon ('scientific') methods that are largely based on logical negotiations between data and theory and that such systematic and relatively unbiased approaches can achieve truths about phenomena of the world. Such positions may lead them to

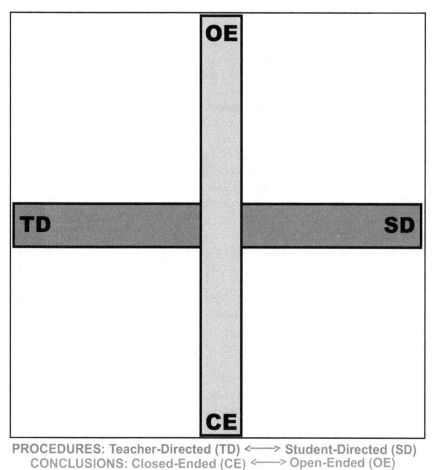

PROCEDURES: Teacher-Directed (TD) <——> Student-Directed (SD)
CONCLUSIONS: Closed-Ended (CE) <——> Open-Ended (OE)

Figure 3.3 Learning control model (Lock, 1990)

greatly control student learning procedures (TD) and encourage them to draw pre-determined (CE) conclusions. Lessons might involve technology-aided (e.g., PowerPoint™) and teacher-guided presentations, with students asking clarifying questions. Students may have opportunities to deepen their understanding of knowledge and skills taught through *guided* inquiry practical ('lab') activities. Teachers may feel comfortable with such approaches because they are relatively easy to plan (often provided in books, etc.) and to assess and evaluate (assuming pre-specified learning goals). Such learning environments also may be comforting to students, administrators, parents and others – perhaps portraying science (and, often, technology/engineering and mathematics) as relatively systematic, unbiased and generally unproblematic in terms of possible adverse influences of its methods and products on societies and environments. Positive portrayals of these sorts may, in turn, encourage students (with parental and other influences) to choose further education – and, eventually, careers – in fields involving science and technology. So to speak, this would make science education seem like an 'infomercial' for professional fields of science and technology.

Extents to which societal members prefer to have fields of science and technology perceived as positive contributors to wellbeing of individuals, societies and/or environments appear quite variable. On one hand, pro-capitalist entities often appear to discredit science – as has apparently been the case with denial of anthropogenic causes of the climate crisis (Klein, 2014) – or at least cast doubts on the veracity of research results that may problematize capitalist products and services like tobacco, nuclear power, pesticides and many more (Oreskes & Conway, 2010; also see: www.merchantsofdoubt.org). At the same time, there is evidence to suggest that pro-capitalist entities have defunded certain fields of science, such as those about climate change, in order to prevent emergence of negative findings about capitalist commodities (Carter et al., 2019). On the other hand, some community members may feel that citizens need full disclosure about possibly harmful products and services associated with science and technology (and related fields) and relationships with people and groups (e.g., capitalists) that may be contributing to such problems. Choosing between extreme *Rationalist-Realist* (RR) vs. *Naturalist-Antirealist* (NA) (Figure 3.1) portrayals of the nature of science and technology can be 'stressful' for teachers. Approaches supporting RR perspectives may encourage students to enrol in future studies and/ or work in these or related fields, for instance, but teachers may feel that some (or many) students may have been misled about their chosen vocations. Having suggested this, some students may – in response to strong, critical, NA portrayals – be motivated to address apparent problems, such as those linked to capitalist influences, in science knowledge development, communication and uses; and accordingly choose further education and possibly careers in these or related fields.

Assuming that most science education programmes err towards more *Rationalist-Realist* depictions of the nature of science (Hodson, 2008), readers of this chapter may find more *Naturalist-Antirealist* approaches described and

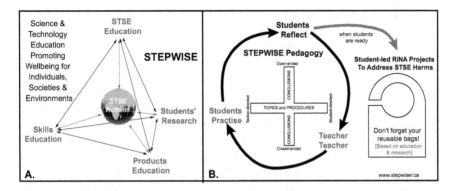

Figure 3.4 STEPWISE theoretical (A) and pedagogical (B) frameworks

illustrated below helpful in debating merits of value-free vs. value-laden portrayals of science and technology. At the same time, however, readers may want to keep in mind Derrida's (1967/1978) suggestion that 'opposites' (e.g., Naturalist vs. Rationalist views) may have unanalyzed merits that may need to be considered.

With assumptions that many or most living and nonliving things seem to have been more or less assimilated into a neoliberal dispositif, which – to a great extent – prioritizes problematic capitalist values like *individualism, competitiveness, commodification, possessiveness* and *cost externalization* (McMurtry, 1999), in 2006 I developed the 'STEPWISE' curriculum framework depicted in Figure 3.4A (Bencze, 2017). STEPWISE is the acronym for *Science and Technology Education Promoting Wellbeing for Individuals, Societies and Environments*. In this tetrahedral schema, five local (Ontario, Canada) curriculum learning 'expectations' (e.g., for *Skills Education*) are shown to be in equal and reciprocal relationships. However, by placing STSE Actions in its geometric centre, the schema is intended to encourage and enable students to 'spend' some of their *cultural capital* (Bourdieu, 1986/2002), such as in terms of their learning in the four peripheral domains (e.g., STSE Education), on *altruistic* social actions that may help overcome harms – like those linked to climate change – in relationships among fields of science and technology and societies and environments (STSE).

The STEPWISE framework is based in part on *actor-network theory* (Latour, 2005), which assumes that all living, nonliving and symbolic entities ('actants') are in networks held together by two-way relationships between and among pairs of actants. With reference to Foucault's (2008) conception of *dispositifs*, however, such relationships can be *biased* – such as when capitalists align myriad actants in support of goals like possessive individualism. *Dispositifs* may, on the other hand, be more *altruistic* – in which actants promote actions that, while perhaps being somewhat egoistic, are greatly aimed at wellbeing for other living and nonliving things (Batson, 2011).

Most teachers who first considered using the STEPWISE theoretical framework (Figure 3.4A) found it to be too complex for arranging lessons and student activities, mainly because it implies that learning in any one domain (e.g., Skills Education) should be simultaneously coordinated with learning in the other four domains. Consequently, the more sequential – 'pedagogical' – schema in Figure 3.4B was developed and judged by several teachers to be more feasible than the tetrahedral model. Nevertheless, most core values of the original tetrahedral framework, such as altruism, were intended to be maintained – although they may always be subject to contextual influences, such as local administrative supports, material resources, teacher beliefs, etc.

The three-phase pedagogy (on the left in Figure 3.4B) is based on *constructivist* learning theory (Osborne & Wittrock, 1985). This theory posits that each person, while being influenced by outside factors like their cultural position, constructs unique attitudes, skills and knowledge (ASK) by combining ASK from experiences with ASK in their brains/bodies from previous experiences. In this light, the schema suggests that lessons begin (in the *Students Reflect* phase) by encouraging students to reflect on and express their current ASK relating to science and technology – often by evaluating, for example, some products of science and technology, such as fast foods when studying human physiology. Because such activities are meant to be mostly student-directed and open-ended (Figure 3.3), students should have freedom to express their 'personal' (although *socially*-constructed) value systems – by saying, for example, that they like fast foods because they make them feel good, or by saying they hate them because of advertising that can trick them into consuming harmful 'food', etc. Encouraging students to express their value systems in such ways can be challenging for some teachers, who may be concerned that such activities use up excessive amounts of time, perhaps limiting opportunities to help students to develop 'appropriate' values, such as those aligned with *Rationalist-Realist* views about science (Figure 3.1).

It may seem appropriate – as argued earlier – to encourage students to express their current ASK prior to teaching about them. Although many of these are likely to be quite varied and, in some cases, quite different from those of mainstream science and technology, such diversities of positions, expertise, etc., are to be *appreciated*. Nevertheless, there may be justification for teaching some, or perhaps most, students about particular attitudes, skills and knowledge, etc., that they may *consider* adopting. Reflecting societies more broadly and their communities more directly, classes of students often feature – to perhaps state the obvious – significant *stratification*. Students vary, of course, in terms of 'abilities' – in part genetically determined, but perhaps also due to nutritional and other differences. A significant contributor to stratification also, however, appears to be sociocultural and economic. Apparently largely resulting from decades of globalized neoliberal perspectives and practices, wealth and wellbeing have increasingly been concentrated into dramatically fewer hands (Piketty, 2020). Increasing fractions of some student populations may consequently have struggled to learn because of deficits in *dominant* cultural and social capital, such as knowledge, speech patterns, abilities to think and work in the abstract and social means for acquiring such benefits (Bourdieu, 1986/2002). This seems to be particularly problematic regarding heavily promoted *inquiry-based learning* approaches, which – although quite varied – can be highly problematic if they are meant to ensure students are expected to 'discover' pre-determined conclusions of mainstream science and technology from experiences with phenomena, searching the Internet, etc. (Bencze & Alsop, 2009). Discovery difficulties may be exacerbated because capitalists apparently often have worked to cast doubts on and/or suppress science findings in public fora (Oreskes & Conway, 2010), and in many STEM education initiatives (Hoeg & Bencze, 2017), that would incriminate their products and services – thus making it difficult for students to discover, and teachers to teach, such problematic information.

To overcome problems of discovery like those mentioned earlier, teachers can *directly* teach about cases (documentaries) that highlight roles of values in STSE relationships and in sociopolitical actions people have taken to overcome related harms of their concern. There are numerous examples of apparent value-related harms in STSE relationships from which teachers can choose. A case pertaining to genetically-engineered (GE) salmon seems very comprehensive and relevant (Pierce, 2015). GE salmon may seem highly attractive as a food source, being much larger than wild salmon. However, they may be like a proverbial Trojan horse; that is, attractive on the 'outside' (e.g., its size) while harbouring potential dangers on the 'inside.' In actor-network theory terms, such commodities are said to be *punctualized* (Callon, 1991); that is, appearing highly simplified (reduced), with people (e.g., consumers) unaware of or unconcerned about its integration into a much larger supportive network (*dispositif*). That network/*dispositif*, although controversial, is suggested by some people to promote highly problematic values – for example, by inflating fish prices and funnelling profits towards company shareholders because of government-sanctioned patents, harming wild and GE salmon because of increased sea lice that flourish in ocean pens and potential unforeseen long-term health harms due to consumption (often without consumer knowledge) of perhaps inadequately-tested GE food. Accompanying documentaries of such controversial commodities should be descriptions of sociopolitical actions that people have taken (e.g., at www.youtube.com/watch?v=oStKYNmGCko&t=14s) to address values-based concerns they have about them. Related to such discussions, teachers may emphasize that effective actions may aim to develop new (or altered) *dispositifs* – with new value systems holding together myriad living and nonliving things, such as educators, activists, new technologies, laws, etc., in support of highly equitable and sustainable food systems. An excellent case in this regard pertains to the different cultural value systems that led South Korea and the USA, respectively, to take different social and technological paths regarding development and uses of nuclear power (Jasanoff & Kim, 2009).

When the teacher feels that students have some initial understandings of potentially problematic STSE relationships and corresponding sociopolitical actions through lessons like those outlined here, students should be asked to *apply* newly taught ASK to different, perhaps related, possibly problematic STSE cases. Such application activities may balance, depending on students' needs, teacher- vs. student-directedness and closed- vs. open-endedness (Figure 3.3). Further examples of such direct teaching, combined with corresponding application activities, are provided through the STEPWISE website (https://wordpress.oise.utoronto. ca/jlbencze/teacher-teaches).

When the teacher believes students have well-developed attitudes, skills and knowledge from the *Teacher Teaches* phase of the STEPWISE pedagogy (Figure 3.4B), students may then be asked to design and carry out practice (*Students Practice* phase) research-informed and negotiated action (RiNA) projects to help overcome harms in STSE relationships of their concern. Such projects are intended to shift balances of student learning control from those that are relatively teacher-directed and closed-ended, as with teacher instruction in the *Teacher Teaches* phase of the pedagogy, through to somewhat more student-directed and open-ended practice RiNA projects, before conceding *much* more control of learning (mostly SD and OE) to students for "Student-led RiNA Projects" (Figure 3.4B). Such transitions in learning control can provide students with a range of interactions with value systems – teaching them about broader societal values, like those linked to capitalism, and encouraging and enabling them to increasingly self-determine values.

Educators have, in terms of the actor-network theory described earlier, myriad influences on their choices. In many or most of our applications of STEPWISE-informed pedagogy, we have erred on the side of strong critiques of influences of capitalist value systems on STSE relationships. Using ideas like those described for the *Teacher Teaches* phase, students have tended to develop sociopolitical actions that challenge capitalist value systems. An excellent example of this is the educational video located at tinyurl.com/3j5mrwc5, in which the student drew on the Trojan horse metaphor, actor-network theory and semiotics to educate people about apparently problematic behind-the-scenes (*punctualized*) pro-capitalist actants associated with liquid foundation makeup, a common commodity. Although we encourage teachers to highlight values supporting *and* opposing capitalism in the *Teacher Teaches* phase, it seems many teachers who have chosen to work with us tend to emphasize problematic aspects of capitalism. Our local network decision to promote critiques of capitalist values may be justified, in part, because pro-capitalist values – although often subconscious in people's minds – tend to dominate societies. Public education may be one of the few places where this may be challenged. Having noted this, we continue to stress to teachers, nevertheless, that in democracies, students must be made to feel free to incorporate their 'personal' (although socially-constructed) value systems into their sociopolitical actions. Omura et al. (2019) suggest that, indeed, people in democracies need to feel free and have resources enabling them to self-determine – often in small, local community groups – value systems they may use to assemble actants to suit their purposes and contexts.

Summary and futures

Teaching, regardless of contexts, involves many values-based decisions – often 'in the spur of the moment' while interacting with learners. In light of apparent hegemonic influences of capitalist individuals and groups on most entities on Earth, including fields of science and technology (or 'STEM'), science educators may consciously or otherwise face ongoing debatable decisions regarding extents to which they might educate students about possibly problematic pro-capitalist relationships among fields of science and technology and societies and environments and, moreover, extents to which they might encourage and enable them to develop and implement plans of action to overcome related harms of their concern. Although discussions in this chapter have undeniably erred on the side of science education that promotes citizenship that is critical of and reformist regarding pro-capitalist systems, they may help science teachers to negotiate – likely involving other educators, administrators, parents, students, curriculum materials, their personal views about science and technology, etc. – their positions regarding education about capitalist value systems in science and technology and beyond.

Questions for further debate

1   Who should have a say in what is taught in science education and how?
2   On what values should science education be based?

Suggested further reading

Barton, A. C. (2001). Capitalism, critical pedagogy, and urban science education: An interview with Peter McLaren. *Journal of Research in Science Teaching, 38*(8), 847–859.

Bazzul, J. (2012). Neoliberal ideology, global capitalism, and science education: Engaging the question of subjectivity. *Cultural Studies of Science Education, 7*, 1001–1020.

References

Batson, D. C. (2011). *Altruism in humans.* Oxford University Press.

Bencze, J. L. (Ed.). (2017). *Science and technology education promoting wellbeing for individuals, societies and environments.* Springer.

Bencze, J. L. (2020). Re-visioning ideological assemblages through de-punctualizing and activist science, mathematics & technology education. *Canadian Journal of Science, Mathematics & Technology Education, 20*(4), 736–749.

Bencze, J. L., & Alsop, S. (2009). A critical and creative inquiry into school science inquiry. In W.-M. Roth & K. Tobin (Eds.), *World of science education: North America* (pp. 27–47). Sense.

Bencze, L., Bowen, M., & Alsop, S. (2006). Teachers' tendencies to promote student-led science projects: Associations with their views about science. *Science Education, 90*(3), 400–419.

Bourdieu, P. (2002). The forms of capital. In N. W. Biggart (Ed.), *Readings in economic sociology* (pp. 280–291). Blackwell. (Original work published 1986)

Cahill, D., Cooper, M., Konings, M., & Primrose, D. (2018). *The SAGE handbook of neoliberalism*. SAGE.

Callon, M. (1991). Techno-economic networks and irreversibility. In J. Law (Ed.), *A sociology of monsters: Essays on power, technology and domination* (pp. 132–161). Routledge.

Carter, J., Berman, E., Desikan, A., Johnson, C., & Goldman, G. (2019). *The state of science in the Trump era: Damage done, lessons learned, and a path to progress*. Union of Concerned Scientists.

Creswell, J. W., & Poth, C. N. (2018). *Qualitative inquiry & research design: Choosing among five approaches* (4th ed.). SAGE.

Davis, M. (2020). *The monster enters: COVID-19, avian flu and the plagues of capitalism*. OR Books.

Derrida, J. (1978). *Writing and difference* (trans. A. Bass). Routledge. (Original work published 1967)

Dzisah, J. (2007). Institutional transformations in the regime of knowledge production: The university as a catalyst for the science-based knowledge economy. *Asian Journal of Social Science, 35*(1), 126–140.

Feyerabend, P. K. (1975). *Against method: Outline of an anarchistic theory of knowledge*. New Left Books.

Foucault, M. (2008). *The birth of biopolitics: Lectures at the Collège de France 1978–1979* (Ed.: M. Sennelart; General Eds.: F. Ewald & A. Fontana; Trans.: G. Burchell). Palgrave Macmillan.

Grant, J. (2018). *Corrupted science: Fraud, ideology and politics in science* (2nd ed.). AK Press.

Hodson, D. (1998). *Teaching and learning science: Towards a personalized approach*. McGraw-Hill Education (UK).

Hodson, D. (2008). *Towards scientific literacy: A teachers' guide to the history, philosophy and sociology of science*. Sense.

Hoeg, D., & Bencze, L. (2017). Values underpinning STEM education in the USA: An analysis of the next generation science standards. *Science Education, 101*(2), 278–301.

Jasanoff, S., & Kim, S.-H. (2009). Containing the atom: Sociotechnical imaginaries and nuclear power in the United States and South Korea. *Minerva, 47*(2), 119–146.

Klein, N. (2014). *This changes everything: Capitalism and the climate*. Simon & Schuster.

Krimsky, S. (2019). *Conflicts of interest in science: How corporate-funded academic research can threaten public health*. Simon & Schuster.

Kuhn, T. S. (1970). *The structure of scientific revolutions* (2nd ed.). University of Chicago Press. (Original work published 1962)

Latour, B. (2005). *Reassembling the social: An introduction to actor-network-theory*. Oxford University Press.

Latour, B., & Woolgar, S. (1986). *Laboratory life: The social construction of scientific facts* (2nd ed.). Princeton University Press. (Original work published 1979)

Li, Y., Wang, K., Xiao, Y., & Froyd, J. E. (2020). Research and trends in STEM education: A systematic review of journal publications. *International Journal of STEM Education, 7*(11), 1–16.

Lock, R. (1990). Open-ended, problem-solving investigations – What do we mean and how can we use them? *School Science Review, 71*(256), 63–72.

Losee, J. (2001). *A historical introduction to the philosophy of science* (4th ed.). Oxford University Press.

Loving, C. C. (1991). The scientific theory profile: A philosophy of science model for science teachers. *Journal of Research in Science Teaching, 28*(9), 823–838.

McGinn, M. K., & Roth, W. M. (1999). Preparing students for competent scientific practice: Implications of recent research in science and technology studies. *Educational Researcher, 28*(3), 14–24.

McMurtry, J. (1999). *The cancer stage of capitalism.* Pluto.

Merton, R. K. (1973). *The sociology of science: Theoretical and empirical investigations.* University of Chicago Press.

Mirowski, P. (2011). *Science-mart: Privatizing American science.* Harvard University Press.

Omura, K., Otsuki, G. J., Satsuka, S., & Morita, A. (Eds.). (2019). *The world multiple: The quotidian politics of knowing and generating entangled worlds.* Routledge.

Ord, T. (2020). *The precipice: Existential risk and the future of humanity.* Bloomsbury.

Oreskes, N., & Conway, E. (2010). *Merchants of doubt.* Bloomsbury Press.

Osborne, R., & Wittrock, M. (1985). The generative learning model and its implications for science education. *Studies in Science Education, 12*(1), 59–87.

Pierce, C. (2015). Learning about a fish from an ANT: Actor network theory and science education in the postgenomic era. *Cultural Studies of Science Education, 10*(1), 83–107.

Piketty, T. (2020). *Capital and ideology* (trans. A. Goldhammer). Harvard University Press.

Wittgenstein, L. (1958). *Philosophical investigations* (3rd ed.) (trans. G.E.M. Anscombe). Macmillan.

Ziman, J. (2000). *Real science: What it is, and what it means.* Cambridge University Press.

Chapter 4

Debates, intimacies, affects and agencies

Science education in the 'hard' climate change era

Steve Alsop

Introduction

Centuries of unabated capitalist-driven industrialisation has wrought lasting changes to the Earth. Since the Great Acceleration, there has been an exponential surge in human impact, resulting in shifts of unprecedented scale, including mass species loss and biodiversity decline, human displacement and ever-increasing disparities between rich and poor. Around 34 billion tons of anthropogenic carbon are now pumped into the atmosphere every year, and a 1.2-degree C increase in global temperature since preindustrial times has already been reached and is being exceeded (IPCC, 2018). The last seven years have been the warmest on record, and debate is steadily shifting to how to keep temperature changes below 2 degrees Celsius; many have accepted that the 1.5C pathway is becoming rapidly unattainable (Tollefson, 2019). The catastrophic impacts of seemingly small shifts in temperature are slowly unfolding, being recorded, analysed, suffered and endured throughout the planet. Rebecca Solnit (2014, np) implores us to call climate change what it is: "violence against places and species as well as against human beings". Greta Thunberg (2019, np), the famous young Swedish climate activist, asks us to "act as though our house is on fire". The Earth Charter opens, declaring: "We stand at a critical moment in Earth's history, a time when humanity must choose its future[1]".

I write this introduction at another critical moment: the Covid-19 pandemic. This continued crisis-upon-crisis has brought civil society to its knees, unnecessarily claiming millions of lives, and in so doing, has laid bare entrenched and tragic local and global inequalities. Those living in poorer countries and in poverty in richer countries disproportionately carry burdens of vulnerable, precarious and fragile living conditions – conditions deeply entwined with histories of marginalisation, economic exploitation, colonialism, racism, displacement, conflict and war. The impacts of climate change – and the compounding coronavirus pandemic – are distributed unevenly and unjustly. Communities bearing the weight of changes (more often women, children, racialised and indigenous communities, poor communities, people with disabilities) are those least responsible, underscoring the significance of recognising the climate crisis as an intersectional, intergenerational crisis of injustice.

DOI: 10.4324/9781003137894-5

This introduction is well known, and, in some way, risks coming across as theatrical and alarmist. After all, there is a long-established history of turning to education to solve global problems, which serve as familiar points of departure for more hopeful and harmonious shores. Clearly, such rhetorical gestures are demonstrable overstatements: education alone cannot end world poverty or solve the climate and biodiversity crisis. The weight of the world cannot, and should not, be placed on the shoulders of teachers and students, especially in ways that abdicate powerful others of their crimes, irresponsibilities and moral duties. But between the reverence and daily practical constraints and realities, I believe there are diverse and rich opportunities for innovative and progressive pedagogies in which science teachers and students come together naming the world and exploring their places in imaginative, life-sustaining and enriching world-building. In the current climate emergency, it is insufficient and inadequate in even the most vibrant and liberal educational forms to lose sight of this possibility and tumble into functionality, despair and day-to-dayness, losing grip of possibilities of joining others – including so many youth-led climate coalitions – with something meaningful, significant and powerful to say and do.

Science education is not presently given the attention that it warrants in regional, national and international climate change discussions and popular mitigation and adaptation strategies (in comparison with, say, green engineering or green economics). Education is only mentioned in passing by the IPCC, with a teacher's guide being made available a couple of years ago (OCE, 2018). It did feature more prominently in COP 26 in Glasgow (in November 2021); however, in past meetings, it has never been a central mitigation or adaptation feature. Nevertheless, in many educational jurisdictions around the world, climate change education is steadily emerging as a science education curriculum priority.

It is perhaps revealing to note that in the previous edition of this book, 'climate change' is only listed twice in the index. Ironically, even I have more mentions, which is a pause for thought (if not lasting anxiety on my part). But over the intervening years, a multiplicity of climate change education-related resources and research projects now exist, and more are emerging (see for example Alsop et al., 2015; Shepardson et al., 2017; Le, 2021).

There is much evidence to suggest, however, that things are not where they should be, given the precipice on which humanity is now perched. A few examples serve to underscore this point. I live in Canada – a country which is warming twice as fast as the rest of the world (BBC News, 2019). Climate change courses currently make up less than 3% of courses in universities (Martinez, 2018), and similar figures are found in other countries. The Ontario curriculum now has a single unit on climate change in Grade 10, although I am frequently reminded of teachers who move this unit to the end of the school year, covering it in a hurried, low-profile, cursory fashion. In the UK, for instance, the 'teach the future' petition-campaign calls for curriculum reforms, teacher education and funding to ensure all students are taught about the climate emergency.[2]

Given the magnitude of the stakes, it is important to ask why changes have been so hesitant and relaxed. What are we waiting for? What is stopping us? There are lots of different reasons in different contexts and situations. Over preceding decades, there has been a loss of autonomy in the teaching profession as state-sanctioned curricula have taken hold. These are slow to change and often conservative in outlook. More is also being asked of teachers than ever before, geared to enhancing efficacy and increasing public accountability. As a consequence, teachers find themselves pulled in numerous directions, uncertain of to whom they are responsible, as well to what, and where. Many teachers continue to grapple with how do I do what I know is right? At times this can seem like an uphill battle, perhaps best directed toward an extracurricular activity or after-school environmental club, rather than reforming and reimagining the curricula mainstream.

It is also important to recognise that teaching climate change is a fraught endeavour. The climate crisis is complex, ubiquitous, difficult to define, 'wicked', politically controversial and emotionally alarming. Given this, it is perhaps unsurprising that it is difficult and disorientating to teach and learn. It doesn't fit easily and, in many ways, should not be forced to fit within the standard school curriculum or popular science teaching methods. As Wilson (2017) playfully asks: "will the end of the world be on the final examination?" Klein (2014) comments that climate change "changes everything". This includes what it means to be a scientist, science teacher, science student and climate citizen, leaving us unsure of who we are, where we are and where we should be going.

Nevertheless, over the next few decades, as the world enters into a period of inescapable and unavoidable change, we will need to come to terms with escalating ruptures in our scientific and education identities, worldviews and practices. So many of us have been brought up and educated in a Western influenced world defined by scientific and technological progress, consumerism, liberalism and seemingly limitless planetary boundaries. These narratives are deeply entwined with dominant educational practices. They are becoming increasingly unstable and untenable. As Greta Thunberg and Autumn Peltier and thousands of students in youth climate movements (including Fridays for Future, EcoGirls and the Indigenous Youth Climate Network) continue to underscore, there is a gaping chasm between what children learn in school, the disastrous effects of global heating and commensurate political (in)actions at multiple scales.

The climate debate

In a book on debates, it is important to recognise the liberating and restricting powers of them. Debates can offer topics necessary explanatory structures and insight. Debates, however, can drag complex topics into polarising positions, hardening identities and lasting conflicts. For far too long, climate change has become framed through a public debate between those who accept global heating and those who don't. This once-vitriolic ideological sparring match between two sides intent on either winning or losing is now fading. It has left in its heavy wake an oversimplification and polarisation of anthropogenic global heating, which has become reduced to a single 'make it or break it' question: do human-related carbon emissions threaten the planet?

Getting this question wrong is often equated with ignorance – lacking understanding of the 'hard' facts and muddled, irrational reasoning. The answer is in the question – if only people understood the science properly, then they would agree on this issue with urgency, intentionality and action. There is a steady flow of studies supporting this common line of reasoning, starting by highlighting particular gaps or deficits in climatic understandings of various identified groups, including the general public, as well as school students and teachers (Hoffman, 2015). While these studies enthral opinion-poll researchers and make for sensational news headlines, the approach conflates important differences between knowledges, knowing, values, affects, contexts and actions.

Understanding scientific knowledge of global heating is clearly fundamentally important (including atmospheric, hydrospheric, cryospheric, lithospheric subsystems, the biosphere, radiative and human forcing and associated climate modelling). Indeed, climate science is far more complex, fluid and far-reaching than the climate debate readily implies, or perhaps publicly allows. But the climate debate is much more than a knowledge gap. As Jassanoff (2010, p. 240) stresses, it is important to recognise that climate facts are "dropped into contexts that have already been conditioned to produce distinctive cultural responses to scientific claims". Important distinctions now need to be made between climate knowledge, climate acceptance and climate denial, highlighting intentionality and orientation – deeply social, relational and affective features of climate cultures and worldviews (see discussions in Hoffman, 2015).

This argument, of course, is far from new. It has been persuasively made over the past few decades. Moreover, the climate debate itself has now changed. As Heron and Dean (2020) point out, there is now widespread acceptance that carbon emissions are affecting the climate – albeit with some high-profile exceptions, including the increasingly out-of-place political leaders ex-President Trump and President Bolsonaro.

Even disinformation campaigns funded by obscenely rich and profoundly destructive fossil fuel companies and their allies have become much more nuanced in their tactics (Rustin, 2021). These often rely on an unquestioned sense of ignorance as individual misunderstanding, masking active 'ignorance creation' and 'doubt mongering' by corporate powers selectively disseminating, manipulating and obfuscating information in pursuit of self-serving economic desires (Milman, 2021). Ignorance, or agnotology, is rarely as logical, personal, apolitical or neutral as schools continue to convey and underscore. Oulton et al. (2004), for example, stress a need for science teachers to embrace knowledge with values and politics when exploring controversial issues. They draw attention to considerable shortcomings of approaches that stick to facts, offer balanced viewpoints and pretend to be politically neutral, proposing an alternative teaching model exploring authority claims and actions of particular groups embroiled in controversy.

This is part of a trend in innovative teaching methods associated with the labels SSI (socio-scientific issues), SAQ (socially acute questions) and STSE (science, technology, society and environment). These attend to ethical and sociopolitical features of complex science-related issues. The ways in which some voices have been historically marginalised in these issues has been a topic of focus – see for example 'The People's Curriculum for the Earth' (Bigelow & Swinehart, 2015).

These approaches join broader critical analyses of ways in which societal 'issues' become de-politicised and individualised, framed as problems to be fixed through technological solutions. Clearly, technology has an important role to play in response, but climate change cannot be addressed without commensurate attention given to human agencies, underpinning causal reasons and social power differentials. As Selby and Kagawa (2010, p. 42) note, the climate crisis involves confronting the "pathology of an ethically numb, inequitable and denatured human condition". The climate crisis challenges traditions of separating culture and nature, and social progress as technological mastery.

Anyhow, much evidence now suggests that the old-school climate divisions have broken down, with new ones now starting to take political form and substance (Heron & Dean, op. cit.). The contemporary question is perhaps no longer if climate change is taking place and should be a defining 'issue' of this era, but rather how climate changes define, or should define, this era, including who we are, what we can do and what futures we might imagine, share and build with each other and all other inhabitants of the planet.

While science education has, in general, been slow to respond to the climate crisis, it has become consumed, for a large part, with an attempt to settle the climate question as a pedagogical solution to a factual reasoning gap. In this regard, climate science education often comes from a deficit perspective of correcting an untruth, rather than inquiry, discovery, releasing climate imaginations or joining with others and acting for political change. This is especially concerning as evidence suggests that increased knowledge of climate science by itself has little or no correlation with pro-climate behaviours (Dijkstra & Goedhart, 2012). Going forward, we really need to urgently redress this balance, recognising that climate science education is less about teaching whether anthropogenic global heating is happening or not, and more about what it means and ought to mean to be a science teacher and science student living in the era of climate change. The climate crisis, in this manner, transforms the very conditions of science, education, and personal and public life. Indeed, public life is perhaps even too restrictive here. As climate science makes demonstrably clear, humans are a living part of the Earth (and climate), and with all living and non-living carbon-producing and exchanging entities, deeply entangled in earthly cycles and systems in flux and imbalance. There are lively contemporary discussions of how we can learn to be more connected, become more 'Earthy', recognising ourselves as agents in planetary systems (rather than passengers) and feeling associated responsibilities towards multispecies flourishing and nourishments. Verlie (2017), for example, advocates teaching climate as a verb, to 'climate', something that humans do together with all other carbon species, rather than as a noun – an object of external human study. Watt-Cloutier (2018) writes that it is "time to listen to Inuit on Climate Change". The much-cited Inuit concept of Sila combines weather, breath and spirit as life-force passing through all living beings. Such examples invite reconsidering underpinning assumptions, language and nomenclature, within which climate gains human sense and worth, propelling dominant cultures forward.

All this is, needless to say, pedagogically complex and far-reaching. At the very least, the challenge here is to move away from a model of teaching in which change happens to children through education (in response to knowledge deficits), rather than education itself being subject to change.

In search of granular intimacy

Norgaard's (2011) study of highly educated adults in Norway shows that while climate change is often well known in the abstract, it often remains steadfastly detached from everyday personal, social and political life. It is simultaneously common knowledge while socially and ecologically unimaginable. Norgaard describes this as a form of "social denial", emblematic of how many citizens in wealthy worlds are responding to global warming. Lack of progress made in science education might be similarly conceived: we know that it is happening, but struggle with connecting this to educational practices and everyday life-worlds for a host of less obvious reasons.

Norgaard's study offers a number of pedagogical challenges to science educators, including how best to connect impersonal, detached globalised climate science with everyday subjective, situated, experiential sense-making. In this respect, the climate crisis invites critical reflection on how to more purposefully and meaningfully respond to temporal, scalar and experiential disjunctions of global/local and disembodied/embodied natures.

Perhaps a key question here is in what ways might teachers and learners form granular everyday intimacy within what are formidable global ecological processes? For many of us, our local experiences of climate changes are small and undramatic: noticing seasonal changes in temperature and weather, hotter summer days and more intense storms, changes in availability of seasonal foods; appearance and disappearance of plants, birds and insects; distant forest fires and murky smoke-filled skies with intense sunsets. This leaves a challenge of how to connect with these encounters, recognising changes as part of global transformations.

Science education can powerfully recalibrate our senses to notice, explore and discuss these changes. Otherwise, this can remain imperceptible, escaping our attentions, empathies, cares and actions. There are a host of wonderful place-based and citizen science monitoring activities in which teachers and students observe and record fluctuating patterns in backyard ice rinks, migrating birds, maple tree sap, local tree species (thriving and declining), exotic and invasive species, including the emerald ash borer and the mountain pine beetle devastating so many northern, lodgepole-jack pine forests in Canada (and across the globe). This is encountering climate change in smaller, granular and intimate ways. Taking up an invitation to gather these data is to re-attend to the mundane and seemingly static, to document patterns, and then perhaps make these patterns public by joining with others and taking action. This is one way of pulling abstract global processes down into meaning-making encounters that can be coupled with actions, simultaneously nurturing attachment, empathy and participatory actions with prospects of empowering teachers, students and communities. The key here is to be touched by what presents itself to the senses, bear witness to what is thriving, fading and lost. More broadly, this is a way of opening into Earthly systems, becoming part of these systems and discussing what responsibilities and adaptations this entails. This can be extended; through media, video and satellite data, it is now possible to attend to more distance and larger-scale changes in similar ways.

Tree planting, monitoring and stewardship has become an extremely popular school activity, accompanied by carbon sequestration calculations and talk of natural carbon solutions. Here it is important to stress ongoing stewardship as a participatory process (not a task), not least because so many trees die within a few months or a year of planting. It is also important to confront timescales involved in drawing down atmospheric carbon through arborescent actions, which offer often longer-term, rather than immediate, carbon mitigation possibilities.

In Canada (and many other countries), there is sustained attention being given to indigenous knowledges and cultural practices, with a host of innovative activities seeking to bring together science and indigenous knowledge (sometimes referred to Native Science). Land-based practices and pedagogies often offer wisdom through close contact, direct sensory interactions with land and spirit, understanding the limits of what can be supported and what is life-sustaining and not. Here, having relationships with and responsibilities for the environment are conceived simultaneously as acts of self-care.

Robin Wall Kimmerer (2013) offers a provocation of walking with Sweet Grass, Linnaeus and Nanabozho as an invitation to braid wisdoms of plants, science and indigenous (Potawatomi) knowledges. In a recent interview conducted during the Covid-19 pandemic (Yeh, 2020), she exclaims: "It is time to take a lesson from mosses", continuing: "what is it that enables them to persist for 350m years, through every kind of catastrophe, every climate change that's ever happened on this planet?" Wall Kimmerer concludes that we need to learn lessons of "being small, of giving more than you take, of working with natural law, sticking together". What a delightful opening this is for science teachers and students to inquire into more modest, ecologically generous and respectful ways of living within local communities and contexts. As science teachers and students, how can we learn to listen to moss, recognising the ecological "gifts" (to use Wall Kimmerer's term) that neighbourhood bryophyta provide?

There is an associated privilege in encountering small and manageable changes. An accompanying activity is to also encounter carbon entitlement, which often remains invisible and obscured by daily comforts and familiar ways of living with nature. There are lots of creative science activities measuring and calculating carbon footprints – thereby exposing and confronting profligate carbon lifestyles and transnational business practices which are now palpably untenable. This is part of a process of reconsidering Western lifestyles and associated convictions that play out in our day-to-day lives.

It is so important to be aware of the collective climate 'we' that analyses can passively assume. For many, including many science students and teachers, climate changes are anything but small. Many are living and working on front lines; struggling to find clean water in droughts and floods; making do with ever-decreasing supplies of food, shelter and firewood; coping with outbreaks of cholera, dysentery and malaria; living on refugee teachers' wages and working in overcrowded and resource-impoverished classrooms. The number of people forcefully displaced through climate disasters now far exceeds those displaced through war and conflict. Refugee organisations record that last year alone, there were 30 million new displacements due to floods, storms and wildfires.[3] In Canada, those most geographically impacted are often living in remote northern and coastal communities. Once more, many indigenous communities are disproportionally impacted, adapting with resilience and fortitude. These communities carry painful histories of apocalypse and genocide wrought through physical and cultural violence of settler colonisation.

Around the world, many science teachers are welcoming climate refugees into their classrooms, listening to and learning from their experiences, offering support, care, new hopes and fresh starts. Many are inviting Indigenous Elders into their classrooms, learning from their intergenerational wisdoms, trauma, relational knowledges and creation stories. Climate testimonies offer more intimate ways of confronting and registering the magnitude and force of this crisis. These stories can now be accompanied by fine-grained, neighbourhood-sized climate data for explorations of regional differences, environmental privileges and injustices, and more harmonious multi-species living. The climate crisis calls upon science teachers and students to search for more equitable ways of living within environmental limits, offering insight and direction for learners to reassess comforts, nurture self-respect, and embrace new responsibilities, participations and possibilities that this unprecedented moment provides.

Embracing climate emotions

There is good reason to be afraid of climate change. News media make this painfully and disturbingly clear. The "slow violence" (Nixon, 2011) of climate collapse and corollaries, including mass extinction and pollution, often lack the impact of the end-of-the-world spectacle, but they are painfully and steadily advancing and being endured.

This dimension of climate crisis is as much to do with sentiment as it is to do with content. Climate change is wearing and demoralising, even for those of us cocooned by privilege. The crisis is constituted with ugly, painful feelings and imagery. There are no shortages of images designed to convey what is at stake. Gore's rising lift scene, in the documentary *An Inconvenient Truth*, dramatically underscores the steepness of the red-lined atmospheric carbon dioxide graph (the hockey stick curve); Paul Walde's installation and sound performance is a moving, solemn requiem for the retreating of Jumbo Glacier in British Columbia, and Andi Snear Magnaso's melancholy letter to the future takes narrative form of a memorial to the lost Icelandic Okjokull glacier, which no longer qualifies as a glacier due to its reduced size. Such examples are just part of a dominating environmental apocalyptic trope. A typical iconography of climate 'disaster porn' includes polar bears precariously balancing on shrinking ice sheets and painfully starving through loss of hunting grounds (Alsop, 2019), and more recently, adorable koalas innocently caught up in scorching and unrelenting Australian bushfires.

Children are growing up and seeking to come to terms with these uncertain, precarious times – times in which images of climate change and the Covid-19 pandemic are part of their everyday lives, their discussions with friends, parents, teachers and community members. Internet-based social media, film-based media, new-media all carry climate (and Covid updates) on an hourly, daily and weekly basis, to the point of near super-saturation. These images are an inseparable part of climate change meaning-making, shaping how people know, imagine, make sense of and relate to the present and future and possible intervention strategies.

Affectively charged encounters are central to this crisis. By breaking with time, dystopian visions offer insight into possible futures, and in doing so they shape visceral senses of urgency to protect the wonders of the natural world, as well as the future of humanity. These are opportunities to feel distant phenomenon through "pedagogical companions" (Alsop, 2019) such as polar bears and koalas, rendering abstract changes to sensory and emotional registers and imaginations (in ways that Keeling curves and atmospheric carbon concentration data simply cannot).

Doomsday representations are, however, not without significant weaknesses. It is an obvious point, but visions set in a period after catastrophe say little about the period leading up to this point. So, rather than fixating on the end of the world, as many have suggested, we need to invite students to imagine possibilities of more livable, more loveable, even more wonderous and fulfilling future worlds. Hofstetter (2019, p. 4) reminds us that we presently lack stories that help us "imagine what a carbon-managed world could look like". Science teachers and students are so well-positioned to draw on their energies and creative capacities to design more radical, visionary and hopeful futures.

Children are themselves emotive symbols of climate futures, anchoring adults' dreams of a future open to change, often through appropriate education. In films and media, children are presented as innocent victims, tragically caught up in adults' monstrous worlds. Such representations invite reflections of how we as adults might respond to children and youth that does not reproduce their innocence and the age-old notion that children and youth need saving, nor leave younger generations to their own devices and futures, or as saviours for adults' reckless follies.

When confronting controversial and upsetting topics, there is always a pressing question of how depressing or hopeful should we be, or can we be? What can we bear? Or perhaps more significantly, what are, can, ought children to bear? Responses, of course, need to be age-sensitive and developmentally responsive. They are also context and situation dependent; I am thinking here of those growing up on climate front lines without the luxuries of choice or developmental shelter.

Climate grief has emerged a topic of extended study, sometimes associated with a series of stages: anger, bargaining, depression and acceptance (Running, 2007). Pihkala (2020) explains – as part of a new BBC series on climate emotions – that climate grief refers to general anxieties, losses and uncertainties associated with climate change. The 'Great Grief" has been a prolonged focus of Albrecht (2019, pp. 38–39), who proposes a new lexicon for 'Earth Emotions', including the terms 'Solastagia' (the pain or distress caused by the ongoing loss of one's home or territory, and 'Terrafurie' (Earth rage). These add to other 'psychoterratic' states that Albrecht has introduced and defined over preceding years.

Such examples at the very least underscore the need for climate education to move beyond cognition and knowledge-lacking approaches and engage learners with/through affect. As factual, rational and 'scientific' as we wish to be, emotions are an inseparable and inescapable feature of the climate era. Climate science education can no longer afford to restrict itself to comforts and confines of facts. This is an invitation for science educators to pay closer attention to affect. As it currently stands, affect is largely framed as a lesser, intermediatory stage to cognition. Associated emotional management approaches commonly seek to 'reduce' or 'overcome' emotions. Here, I think we should be wary of deficit models which focus on getting emotions 'right', which usually means ignoring them or seeking to rationalise them away. Research on emotions in the teaching and learning of socio-scientific issues and controversies is scant. The focus has been placed on values, ethics and politics (to a lesser degree) in ways that underscore rationalities and logicalities, rather than their affectations. This is clearly a critical area for future research and practice to explore.

An associated challenge is how to meet climate emotions as *both* motivating and demotivating forces (avoiding the familiar impulse of predetermining emotions as either good *or* bad). Despair, guilt, anger and frustration can be paralysing, undermining the self in ways that are debilitating and weakening. But these same emotions can also be empowering and energising. Stoknes (2015, p. 183), for example, writes of the "despair paradox" – that the "more that we let death in, the more we can appreciate life's vibrancies". There is much evidence that those moved by apocalyptic images have experienced personal transformations, been inspired to act, and brought about truly necessary and lasting changes.

There are, perhaps, few formulaic and generalisable answers here for educators. One widely circulating recommendation, however, is to embrace emotions by sharing them and talking about them. Albrecht (op. cit.), Pihkala (op. cit.), and many others stress that by naming emotions they can be reconsidered personally and shared socially. Climate mourning is an associated process linked with talking and reflecting on losses and relearning, coming to terms with a different world, confronting its unpredictable and uncertain natures. Fear, anger and frustration are a complement of emotions readily associated with the climate crisis. These can be volatile and simultaneously enriching and emboldening, constituting the raw fabric within which choices and commitments are forged, values and life trajectories become defined, and actions taken. At such intense moments, perhaps threats of indifference and apathy reveal themselves, calling upon our critical attentions. At the very least, science teachers and students might offer opportunities for sharing anxieties and feelings, coming together, supporting each other at this extraordinary terrestrial moment.

Wilson (2017) writes of his experiences teaching a first-year undergraduate course on climate change. He arrived anticipating a soul-searching emotional ride but was confronted with students' 'indifference'. There is always a tendency to think of emotions as individualised psychologised states; however, Norgaard's (2011) study demonstrates ways that climate change is framed in the neoliberal public sphere to provoke feelings of guilt, fear and hopelessness, promoting indifference. There is guilt from individual carbon footprints that are too large; fear that the ecological crisis is so immense and globalised; and hopelessness because it seems impossible to do anything meaningful about it. In response, Wilson's curriculum focuses on *anger* (through readings and discussion of the collective failures of governments, and the ways that this crisis is often projected onto individuals, making them feel guilty), *entitlement* (as a reaction to media discussions of youth as entitled, Wilson includes readings of climate entitlement, underscoring ways that the stable climate that has supported humanity for the past 10,000 years) and *empowerment* (readings of the growth and power of social justice and environmental movements).

A very popular theme of contemporary discussion is the relationships between climate emotions, agencies and actions. Luisa Neubauer, a high-profile German youth activist, offers advice on how young people can tackle the climate crisis (Young-Powell, 2021, np). It starts by "allowing yourself to be touched by what we're seeing around the world. Feel grief [at what's already been lost] and joy about what's still there". Then "turn this anxiety into something constructive and empowering" – "join a movement" and "take to the streets", "push for systemic change", and "learn from the pandemic [which has shown] we can listen to the science". As Greta Thunberg reflects, the cause of her anxiety is not with learning facts; it is with those who have agency who are not listening or doing anything with these facts (see discussions in Kouppanou, 2020).

Cultivating student agency

Young people have been profoundly impacted by compounding crises of the pandemic, systemic racism and climate change. Nevertheless, their resolve and resilience has been sustained, measured and steadfast, as evidenced in the diversification and proliferation of youth-led social and environmental justice organisations. Indeed, the last three decades have witnessed paradigmatic transformations in how youth engage in politics and advocacy. Well-established environmental organisations, such as Greenpeace and The World Wildlife Fund, have been joined by a vast array of youth-led social movements and advocacy groups, including Fridays for Future, UK Youth Climate Coalition (UKYCC), Sunrise in the US, and, more locally, Toronto Youth Climate Justice. These are made possible by the skills and determination of youth leaders as well as Internet and mobile digital devices, empowering new political landscapes and eschewing traditional lobbying practices. Youth are forging new ways of representing, producing and disseminating climate multimedia and collectively organising with global outreach orientated towards lasting changes.

Teachers are reaching out, attending demonstrations and supporting the youth climate movement. Nevertheless, there is a need for education to catch up with these movements, as Rousell and Cutter-Mackenzie-Knowles (2020, p. 203) note: "rather than shying away from the earth's looming runaway climate change, the learning moment can be seized to think about what really and profoundly matters".

One thing that can be learned from these movements is the significance of creative, participatory and technologically mediated approaches, which offer opportunities for young people to engage with the climate crisis in ways that are situationally, personally and culturally consequential and empowering. This can take a variety of forms in which science teachers and students work together with youth climate organisations and community organisations and groups in action-orientated, collaborative mitigation and adaptation projects. Such collaborations offer teachers and students opportunities to understand and redress changes in local (and more distant) communities and environments.

Participatory action-orientated educational approaches offer sense-making and liberating ways for engaging children and young people with climate change. Simultaneously, they open up possibilities to reconsider, retheorise and respond differently to youth agencies. Education often rests on an image of youth as lacking knowledge (that familiar knowledge gap), or as citizens-in-progress or scientists-in-the-making. As a consequence, children are often conceived of as in need of change rather than as subjects and partners in change (see discussions in Alsop & Bencze, 2014).

The climate crisis and the youth climate movement position youth in different, more progressive ways, recognising young people as having distinctive climate perspectives and rights – they are not future citizens, but citizens in danger of losing a future. As Klein (2019, p. 7) comments:

> Young people around the world are cracking open the heart of the climate crisis, speaking of a deep longing for a future they thought they had but that is disappearing with each day that adults fail to act on the reality that we are in an emergency.

Greta Thunberg and other youth activists also stress this point – adults are destroying the Earth and thereby denying children of their futures.

This invites reconsidering adult-educator and youth-learner relationships, throwing into question traditional, and still very popular, top-down climate science teaching approaches. Nevertheless, it is important within educational practices to recognise that children are still vulnerable political agents with less experience and expertise, and are rarely granted the same rights as adults automatically have. This points to more innovative teaching which offers youth insights into and opportunities for ways of claiming and exercising their political rights. There has been a steady growth in science education research and practice with both activist (Alsop & Bencze, 2014) and protest (Lowan-Trudeau, 2019) agendas. These approaches support and structure possibilities for youth to bring personal, ethical commitments and scientific expertise to bear on particular situations and recognise their distinctive agencies. These include educational activities to develop subjectivities through learning about successful political strategies and taking action. For the past decade, I have been coediting a journal, *JASTE* (*Journal of Activist Science and Technology Education*[4]), which contains many examples of different approaches from the perspectives of researchers, teachers and students. A central theme of this work is exposing and reconsidering neoliberal convictions that dominate everyday practices and acceptances. Beautiful Trouble[5] is another web-based resource offering creative tools for a more just world.

To give one example: as a regional response to a worldwide youth climate strike (September 2019), Roxy Cohen (a graduate student and youth activist) and I organised and took part in a 'teach-in'. It was, in part, a challenge to high-profile critics who see climate action as an excuse to miss school. The Toronto Fridays for Future chapter proposed the theme of how youth can talk to adults about climate change and be taken more seriously.

On a field outside of the Ontario parliament buildings, more than 400 students (from elementary, secondary and university) joined in a series of rotating workshops focusing on how to talk to family, teachers and politicians. Invited politicians, teachers and parents took part in these workshops. I discussed with groups of students about how to talk to teachers. We explored: what form and topics might such conversations take? What changes do youth want? To whom should these requests be directed? How can youth's opinions be heard? How can they be more effective agents of change? In a wide-ranging discussion, the students and I discussed collaboration, school governance structures, what to ask administrators and how to approach teachers for support. We talked about who and what controls the school day (including the timetable and government policy). How have schools and other institutions mitigated and adapted to climate change? How might schools offer climate data and leadership in local communities? One of the recommendations, made by a 13-year-old student, was for the creation of a youth-led time-tabled lesson on Fridays, called Fridays for Future, in which youth and teachers discuss climate change and possible futures. In response to identified needs of leadership expertise, last year Roxy and I worked with local indigenous, climate justice and racialized youth organisations to develop a university-based course focusing on youth climate leadership, called 'Rooted and Rising[6]'.

Discussions of youth agencies once more cut across the grain of science education research and practice. In science education, they invite reconsidering and redressing what "childness and children's agency mean" (Kouppanou 2020, p. 958) and ought to mean. In the climate era, politics, children, science and education have all collided in new ways, exposing traditions and educational orthodoxies. This has enabled new coalitions to coalesce, exploring ways to democratise climate science – youth groups, scientists, indigenous rights and social justice activists are working together. Even David Attenborough, the iconic nature presenter, has recently gone on record in a parliamentary select committee on environmental policy, stressing that "we cannot be radical enough".[7] The climate crisis has ruptured the public sphere, rendering visible traditions and hierarchies and thereby offering opportunities for more hopeful, more progressive possibilities of democratising climate politics.

Embracing this extraordinary moment

Science education is such an important part of our climate crisis mitigation and adaptation responses because it has the capability of generating new questions, new possibilities, new imaginaries, new openings, new relationships, new agencies and actions. It offers opportunities for teachers and learners to see, sense and talk about climate afresh, to confront despair and indifference by cultivating empathy, hopefulness and joyfulness through actions embracing endless possibilities of more liveable, fulfilling, vibrant multispecies worlds. Such approaches offer ways of expanding the climate science education genre beyond stale knowledge deficits and fatalistic dystopian endings, realising imaginations within granular intimacies, revitalising emotional encounters and inspiring youth actions for lasting change. In this way, perhaps, the climate crisis has the potential to transform and elevate the status of science education as a subject of profound importance for climate mitigation and adaptation.

To end with an existential tone: as science educators, we have become thrust into the climate era and the inescapable responsibilities that this extraordinary moment entails. Past traditions cannot relieve burdens of freedom, but with our students and many other inhabitants of the planet, we need to decide how best to work together for the world that we want, and within these processes accepting responsibilities for this world.

Questions for further debate

1   How can we teach about climate education in ways that incorporate an emotional dimension?
2   To what extent should science teaching acknowledge and include the student voice?

Suggested further reading

Jones, C. A., & Davidson, A. (2021). Disempowering emotions: The role of educational experiences in social responses to climate change. *Geoforum, 118,* 190–200.

Stevenson, K. T., Petersen, M. N., & Bondell, H. D. (2018). Developing a model of climate change behavior among adolescents. *Climatic Change, 151,* 589–603.

Notes

1   See https://earthcharter.org/
2   www.teachthefuture.uk/
3   I am currently teaching in the Dadaab refugee camp in north Kenya as part of a borderless higher education initiative. Listening to the experiences of student teachers has made these conditions feel very close and real. See www.bher.org
4   See https://jps.library.utoronto.ca/index.php/jaste
5   See www.beautifultrouble.org/
6   See www.bonesthrown.com/rootedandrising
7   See www.theguardian.com/tv-and-radio/video/2019/jul/09/we-cannot-be-radical-enough-david-attenborough-climate-emergency-video

References

Albrecht, G. (2019). *Earth emotions: New words for a new world.* Cornell Free Press.

Alsop, S. (2019, August 26–30). *How to care with polar bears and Jacques Ranciere: Exploration of political aesthetics in science education* [Paper presentation]. European Science Education Research Association, Bologna, Italy.

Alsop, S., & Bencze, L. (2014). Activism! Toward a more radical science and technology education. In L. Bencze & S. Alsop (Eds.), *Activist science and technology education* (pp. 1–21). Springer.

Alsop, S., Greenwood, D., Vaughter, P., & Scott, S. (2015). *Climate change education: Acting for change.* PEASE.

BBC News. (2019). *Canada warming twice as fast as the rest of the word, report says.* https://www.bbc.co.uk/news/world-us-canada-47754189

Bigelow, B., & Swinehart, T. (2015). *A people's curriculum for the earth.* A Rethinking Schools Publication.

Dijkstra, E. M., & Goedhart, M. J. (2012). Development and validation of the ACSI: Measuring students' science attitudes, pro-environmental behaviour, climate change attitudes and knowledge. *Environmental Education Research, 18*(6), 733–749.

Heron, K., & Dean, J. (2020). Revolution or ruin. *e-flux Journal #110.* http://worker01.e-flux.com/pdf/article_335242.pdf

Hoffman, A. (2015). *How culture shapes the climate debate.* Stanford University Press.

Hofstetter, D. (2019, February 22). Writing for impact: How climate fiction can make a difference. *Medium.* https://dhofstetter.medium.com/writing-for-impact-how-climate-fiction-can-make-a-difference-e7b27e4453dd

Intergovernmental Panel on Climate Change (IPCC). (2018). *Global Warming of 1.5°C – An IPCC special report on the impacts of global warming of 1.5°C above pre-industrial levels and related global greenhouse gas emission pathways, in the context of strengthening the global response to the threat of climate change, sustainable development, and efforts to eradicate poverty.* IPCC.

Jassanoff, S. (2010). A new climate for society. *Theory, Culture & Society, 27*(2–3), 233–253.

Klein, N. (2014). *This changes everything: Capitalism vs the climate.* Simon and Schuster.

Klein, N. (2019). *On fire: The burning case for a green new deal.* Penguin, Random House.

Kouppanou, A. (2020). Environmental education and children's agency at the time of the Anthropocene. *Journal of Philosophy of Education, 54*(4), 944–959.

Le, K. (2021). *Teaching climate change for grades 6–12.* Routledge.

Lowan-Trudeau, G. (2019). *Protest as pedagogy: Teaching, learning, and Indigenous environmental movements.* Peter Lang.

Martinez, A. (2018). *Climate change education: An exploration of curriculum enactments in Canadian post-secondary institutions.* Dissertation, York University. https://yorkspace.library.yorku.ca/xmlui/handle/10315/34326

Milman, O. (2021). World is failing to make changes needed to avoid climate breakdown, report finds. *The Guardian.* https://www.theguardian.com/environment/2021/oct/28/world-failing-make-changes-avoid-climate-breakdown-report.

Nixon, R. (2011). *Slow violence and the environmentalism of the poor.* Harvard University Press.

Norgaard, K. (2011). *Living in denial: Climate change, emotions and everyday life.* The MIT Press.

Office for Climate Education [OCE]. (2018). *IPC special report. Global warming of 1.5C. Summary for teachers.* Author.

Oulton, C., Dillon, J., & Grace, M. (2004). Reconceptualising the teaching of controversial issues. *International Journal Science Education, 26*(4), 411–423.

Pihkala, P. (2020). *How can we mourn a changing planet, BBC future: Climate emotions.* https://www.bbc.com/future/article/20200402-climate-grief-mourning-loss-due-to-climate-change

Rousell, D., & Cutter-Mackenzie-Knowles, A. (2020). A systematic review of climate change education: giving children and young people a 'voice' and a 'hand' in redressing climate change. *Children's Geographies, 18*(2), 191–208.

Running, S. W. (2007). *The 5 stages of climate grief.* Numerical Terradynamic Simulation Group Publications 173. https://scholarworks.umt.edu/ntsg_pubs/173/

Rustin, S. (2021, May 17). How we talk about the climate crisis is increasingly crucial to tackling it. *The Guardian.* https://www.theguardian.com/commentisfree/2021/may/17/talk-about-climate-crisis-tackling

Selby, D. & Kagawa, F. (2010). *Education and climate change: Living and learning in interesting times.* Routledge.

Shepardson, D., Roychoudhury, A., & Hirsch, A. (2017). *Teaching and learning about climate change.* Routledge.

Solnit, R. (2014, April 7). Call climate change what it is: Violence. *The Guardian.* https://www.theguardian.com/commentisfree/2014/apr/07/climate-change-violence-occupy-earth

Stoknes, P. E. (2015). *What we think about when we try not to think about global warming: Toward a new psychology of climate action.* Chelsea Green Publishing.

Thunberg, G. (2019, January 25). Our house is on fire: Greta Thunberg, 16, urges leaders to act on climate. *The Guardian.* https://www.theguardian.com/environment/2019/jan/25/our-house-is-on-fire-greta-thunberg16-urges-leaders-to-act-on-climate

Tollefson, J. (2019). The hard truths of climate change – by the numbers. *Nature, 573*(7774), 324–328.

Verlie, B. (2017). Rethinking climate education: Climate as entanglement. *Educational Studies, 53*(6), 560–572.

Wall Kimmerer, R. (2013). *Braiding sweetgrass: Indigenous wisdom, scientific knowledge, and the teaching of plants.* Milkweed Editions.

Watt-Cloutier, S. (2018). It's time to listen to the Inuit on climate change. *Centre for International Governance Innovation.* https://www.cigionline.org/articles/its-time-listen-inuit-climate-change/

Wilson, R. (2017). Will the end of the world be on the final exam? Emotions, climate change, and the teaching an introductory environmental studies course. In S. Siperstein, S. Hall, & S. LeMenager (Eds.), *Teaching climate change in the humanities* (pp. 53–59). Routledge.

Yeh, J. (2020, May 23). Robin Wall Kimmerer: 'People can't understand the world as a gift unless someone shows them how'. *The Guardian.* https://www.theguardian.com/books/2020/may/23/robin-wall-kimmerer-people-cant-understand-the-world-as-a-gift-unless-someone-shows-them-how

Young-Powell, A. (2021, January 27). Climate activist Luisa Neubauer: 'How can we turn this anxiety into something constructive?' *The Guardian.* https://www.theguardian.com/environment/2021/jan/27/cimate-activist-luisa-neubauer-how-can-we-turn-this-anxiety-into-something-constructive

Chapter 5

Climate change education

Justin Dillon

Introduction

On the day that I started to write this chapter, I learned that six young people from Portugal were bringing a case against the governments of 33 countries in the European Court of Human Rights. They ranged in age from eight to 20, and their argument was that the governments had failed to respond adequately to climate change (Slingo, 2020). These six are not alone – millions of young people around the world are demanding to be heard. A key part of their argument is that the education system has failed to educate them about climate change. Increasingly, older people are paying attention; indeed, Greta Thunberg, the climate change activist and founder of the Fridays for Future movement, has twice been nominated for the Nobel Peace Prize. However, there are others who are resistant to changing the curriculum. This chapter looks at the debate about the place of climate change in the curriculum and why the issue is so contentious. I begin by looking at the relationship between nature, the environment and science education.

Education and the natural environment

Children were taught about nature and the natural environment long before we had schools and teachers. Adults taught their offspring how to hunt and kill, gather and grow, long before we knew how to read and write. More recently, nature study was a recognised part of the UK primary curriculum during parts of the 20th century, and rural studies were taught in some secondary schools for decades. However, despite the centrality of the environment to our very existence, environmental education has been at the periphery and not at the centre of the curriculum.

People have been advocating for greater inclusion of nature and the environment for hundreds of years. For example, in 1651, Samuel Hartlib wrote the 'Essay for Advancement of Husbandry Learning', which focused on the education of apprentices. In subsequent centuries, a number of independent

DOI: 10.4324/9781003137894-6

schools introduced agriculture into the curriculum so that their students could farm and manage their estates more effectively. Contrasting with this utilitarian rationale is a more philosophical justification. Rousseau's 'Emile', published in 1767, summarised the author's argument that nature teaches children, not teachers. Johann Pestalozzi built on this philosophy in his Swiss schools in the second half of the 18th century. Others to experiment with this approach to pedagogy were Froebel and Herbart in Germany. Despite the recent growth in the number of forest schools, it is still the case that the environment is not as central to people's education experience as it might be, and the consequences are all too visible.

When I was a science teacher in the 1980s, the curriculum encouraged a focus on human impacts on the environment. I had students in my classes simulate oil slicks in petri dishes and then try to find the best way to tackle them using sawdust or detergent. We also looked at acid rain and at how it eroded buildings and statues made of limestone. A few years later, I was teaching A-level students about what caused the hole in the ozone layer and the need to reduce our use of CFCs. Such problems vary in their complexity, but all of them can be solved in one way or another. However, we are now faced with a set of problems that have no immediate or long-term solutions, such as biodiversity loss, poverty and climate change, all of which are classed as 'wicked problems'. Educating people to cope with wicked problems is one of the great challenges facing educators today (Wals et al., 2014).

The case for climate change education in the curriculum

In 1989, the United Nations adopted the Convention on the Rights of the Child. Article 29 of the Convention states that all parties signing up "agree that the education of the child shall be directed to . . . [t]he development of respect for the natural environment". The English national curriculum dates back to the late 1980s, the same time as the UN's convention. A range of subjects were specified that had to be taught and there was a set of cross-curricula themes, one of which was environmental education. It did not take long before a consensus emerged that what was proposed was too prescriptive and too big. Since then, several attempts have been made to revise the curriculum and to slim it down. Each revision provides opportunities for new material to be added or for a shift in focus. At the first major revision in 1994/5, the cross-curricula themes disappeared. However, climate change as a topic did appear.

In 2007, the topics 'cultural understanding of science' and 'applications and implications of science' were added to the curriculum for 11- to 14-year-olds. Teachers were provided with several opportunities to teach about climate change and global warming (Box 5.1).

Opportunities to teach about climate change in the 2007 national curriculum (Shepherd, 2011)

What the national curriculum says children should learn about climate change and caring for the environment in science lessons:

Age 5–11: Pupils should be taught to care for the environment as part of a topic on life processes and living things.

Age 11–14: Pupils should be taught how human activity and natural processes can lead to changes in the environment and about ways in which living things and the environment need to be protected. Teachers are encouraged to examine issues such as the finite resources available to us, waste reduction, recycling, renewable energy and environmental pollution.

Pupils demonstrate exceptional performance if they can "describe and explain the importance of a wide range of applications and implications of science in familiar and unfamiliar contexts, such as addressing problems arising from global climate change".

Age 14–16: Pupils should learn that the surface and the atmosphere of the earth have changed since the earth's origin, and are changing at present. They should also study how the effects of human activity on the environment can be assessed, using living and non-living indicators. Under "applications and implications of science", pupils should be taught to "consider how and why decisions about science and technology are made, including those that raise ethical issues, and about the social, economic and environmental effects of such decisions".

Climate change also comes up in the geography curriculum and may be tackled in religious education, too.

Michael Gove was appointed Education Secretary in 2010 and commissioned a review of the curriculum. Gove argued that the national curriculum was "too long [and] patronising towards teachers and stifled innovation" (Shepherd, 2011). He was also quoted as saying that: "Its pages are littered with irrelevant material – mainly high-sounding aims, such as the requirement to 'challenge injustice', which are wonderful in politicians' speeches, but contribute nothing to helping students deepen their stock of knowledge" (Shepherd, 2011).

Arguments for the inclusion of more material about climate change in the science curriculum were resisted by the man tasked with leading the change, Tim Oates:

In an interview with the *Guardian*, Oates called for the national curriculum "to get back to the science in science". "We have believed that we need to keep the national curriculum up to date with topical issues, but oxidation and gravity don't date," he said. "We are not taking it back 100 years; we are taking it back to the core stuff. The curriculum has become narrowly instrumentalist."

(Shepherd, 2011)

Expanding on his ideas three days later, Oates laid out his argument in a letter to the *Guardian* (June 16, 2011):

There is a vital distinction between what the national curriculum should prescribe as core scientific knowledge and the vast range of issues that can and should be discussed in schools that need not be listed in the national curriculum. . . . My review will give teachers more time and freedom to explore issues that are relevant to their pupils and the modern world. So, we fully expect schools to tackle important issues like climate change as part of their own curriculum. This distinction between the national curriculum and the school curriculum will enable us to raise standards of scientific literacy, ensuring that an increasing range of issues – including climate change – is subject to lively, informed debate in schools. I am not saying certain issues do not need to be taught. But some things do not need to be taught in a prescriptive way, at a set time. Teachers do not need to be told how to do their jobs.

Michael Gove was quoted in the *Times* as saying:

One of the problems we have had with science in the past is that people have said 'in order to make science relevant you've got to link it to things which are contemporary – climate change or food scares' but . . . what they need is a rooting in the basic scientific principles, Newton's laws of thermodynamics and Boyle's law.

(We Need Stretching Exams . . . , 2011)

Oates' and Gove's positions were criticised by a number of commentators, including Professor Bill Scott, founding editor of the journal *Environmental Education Research*. In his blog post entitled 'Studying climate's ok (probably); but studying climate change definitely isn't', Scott wrote:

Curriculum is always a selection from culture, and by its very nature involves a compromise between competing social goals (e.g., cultural transmission (or rejection)/social change (or not)/employment and employability/personal fulfilment and well-being). Wise societies choose carefully and re-think their choices from time to time, especially when faced with social or economic challenges. The need for sustainable development is thought by many to be such a point of choice – though not, I suspect by current ministers.

(Scott, 2011)

Another critic was Bob Ward, policy and communications director of the Grantham Research Institute on Climate Change and the Environment at the London School of Economics, who "warned that Oates' ideas might not be in pupils' best interests and could make science less interesting for children". Ward added:

> An emphasis on climate change in the curriculum connects the core scientific concepts to topical issues. . . . Certain politicians feel that they don't like the concept of climate change. I hope this isn't a sign of a political agenda being exercised.
>
> (Shepherd, 2011)

But it was, as we shall see later. The review itself did not go smoothly. By June 2012, there were reports that Michael Gove was in "open warfare with the panel of experts commissioned to advise the government on the controversial national curriculum" (Boffey, 2012). One of the complaints was that a "shadow team of advisers, whose identity has not been made 'transparent', was advising Gove and bypassing the official panel in a move that raised concerns among the experts" (Boffey, 2012).

The review was finally published in March 2013 and its recommendations opened up for public consultation. The *Guardian* reported that:

> Debate about climate change has been cut out of the national curriculum for children under 14, prompting claims of political interference in the syllabus by the government that has failed "our duty to future generations". . . . The latest draft guidelines for children in key stages 1 to 3 have no mention of climate change under geography teaching and a single reference to how carbon dioxide produced by humans impacts on the climate in the chemistry section.
>
> (Jowitt, 2013)

The *Guardian* went on to claim that both scientists and climate campaigners were alarmed because, they argued, "teaching about climate change in schools has helped mobilise young people to be the most vociferous advocates of action by governments, business and society to tackle the issue" (Jowitt, 2013).

Replying to the consultation, the Open University's Joe Smith argued that:

> Their removal appears political: playing to imagined prejudices of a Tory right that recalls a globe half draped in the Union Jack. It is a melancholy fact that Mr Gove and his Cabinet colleagues are unlikely to experience the worst of the anticipated consequences of the current lack of political vigour on environmental issues. It is the young who will get to fill in the gaps in their geography curriculum first hand across the course of their lives.
>
> (Smith, 2013)

The government's former science adviser, Prof. Sir David King, was quoted as saying: "What you seem to have is a major political interference with the geography syllabus" (Jowitt, 2013). King's argument was that "climate change should be taught alongside the history of – successful – past attempts to curb chlorofluorocarbon (CFC), which is blamed for the depletion of the ozone layer, and air pollution caused by coal fires and cars", adding, "It would be absurd if the issues around environmental pollution weren't core to the curriculum. I think we would be abdicating our duty to future generations if we didn't teach these things in the curriculum" (Jowitt, 2013).

Geographers, however, supported the new guidelines. Both the Geographical Association and the Royal Geographical Society (RGS) expressed their support. Rita Gardner, the director of the RGS, argued that:

> In the past, in some instances, young people were going to start on climate change without really knowing about climate. . . . What we have got [in the new draft] is a much better grounding in geography, and it has the building blocks for a much better understanding of climate change and sustainability.
>
> (Jowitt, 2013)

John Ashton, founder of the independent not-for-profit group E3G, argued that: "The shift of any mention of climate change from geography to chemistry makes me more concerned, not less", adding: "What's important is not so much the chemistry as the impact on the lives of human beings, and the right place for that is geography" (Jowitt, 2013).

The Department for Education rebutted the suggestion that climate change was being excised from the national curriculum: "All children will learn about climate change. It is specifically mentioned in the science curriculum and both climate and weather feature throughout the geography curriculum" (Jowitt, 2013). Similarly, at the Department for Energy and Climate Change, sources were upbeat about the proposed changes: "There's nothing from the DfE that says climate change is off the agenda or will never be taught. Sensible teachers will look at that as the broadest of signposting". This point is, in essence, what Tim Oates was arguing some years earlier. The UK Youth Climate Coalition (UKYCC), however, argued that leaving the decision about whether to teach about climate change to individual teachers risked it not being taught. Camilla Born, an international expert at UKYCC, stated, "It appears climate change is being systematically removed from the curriculum, which is not acceptable when this is the biggest challenge our generation is going to face, the biggest challenge future generations are going to focus on" (Jowitt, 2013).

Sarah Lester, a policy researcher specialising in climate change education at the Grantham Institute of Climate Change at Imperial College, London, also expressed the idea that climate change was being stripped out of the curriculum. She rejected the argument of the RGS, that pupils first needed to learn the 'building blocks' before they were taught about climate change:

The Government's argument for this seems to be that students should be taught the "building blocks" of climate science before learning about human impacts on the climate and environment. However, as the curriculum currently teaches these building blocks within the sciences and within Geography this argument seems inaccurate.

(Lester, 2013)

Lester added:

Regardless of how strongly you feel about climate change, allowing the Government to make such drastic alterations to the National Curriculum without justifying these changes should ring alarm bells. History, languages and the broader science curriculum have also been changed substantially. While these particular changes will only be implemented for under-14-years-olds, the Department for Education is also making changes to the GCSE curriculum. We need to care, not just about climate change and the sustainability of our resource use, but also about how we hold Government to account and what our children are learning for their future.

(Lester, 2013)

Writing in the *Guardian*, Leo Hickman (who went onto be chief adviser for climate change at WWF-UK) argued that the omission of climate change from the new geography curriculum for under 15s "should not be cause for panic" (Hickman, 2013).

I agree with the view that climate change should largely be taught under the umbrella of geography (both human and physical geography). Yes, it reaches across many subjects – including, obviously, the physical sciences – but geography seems the most appropriate subject from which to "discuss" it as a topic. However, in my own view, key stage 3 is the earliest point at which pupils should begin to specifically address climate change in the classroom.

(Hickman, 2013)

Hickman went onto argue that "from what I can see, this is still possible under the new draft guidelines, given the intentionally loose wording (which starkly contrasts with the detailed, prescriptive wording of the previous version)". So, as before, the main argument is that teaching about climate change was "still possible". Alice Bell, however, had no truck with Hickman's line of thinking:

I've seen rhetoric like this before. Apparently looser codes of education often mask less obvious ways of allowing the easy replication of social hierarchies and divides through education systems. If we can collect together to learn to spell (see appendix 1 of the consultation document), we can think about huge global challenges like climate change together too.

(Bell, 2013)

Referring back to Tim Oates' desire two years earlier to "move away from the teaching of scientific "issues" and instead "to get back to the science in science", Bell wrote:

> The idea of getting back to the science sounds good too, doesn't it? Smells refreshingly rigorous. It's also tripe. The scientific equivalent of back to basics (a Conservative slogan from the 1990s), hollowly looking to an imagined space we probably don't really want to return to even if it had ever existed. . . . The way we conceive of science education is often too atomised around abstract disciplines and too focused on building neat divided sets of science undergraduates, rather than serving and connecting the public at large.
>
> The building of any curriculum is a highly political act, a matter of picking and choosing. In many ways it's a collective expression of what we as a nation value. That's why it's so controversial, because we disagree over what's most important. I think climate change is important, and I think the Department for Education should too.
>
> (Bell, 2013)

Back in 2011, Bob Ward had argued that omitting climate change from the national curriculum could open the door to teachers who are climate change sceptics to abandon teaching the topic:

> This would not be in the best interests of pupils. It would be like a creationist teacher not teaching about evolution. Climate change is about science. If you remove the context of scientific concepts, you make it less interesting to children.
>
> (Shepherd, 2011)

Two years later, Ward was arguing that:

> The trouble for Oates and Gove is that they were displaying an utter ignorance of the history of science. The study of climate change is much older than the discovery of plate tectonics or the structure of DNA, for instance, with Svante Arrhenius in 1886 publishing the first calculations of how much global warming will occur as the levels of carbon dioxide in the atmosphere increase.
>
> (Ward, 2013)

Ward noted that there had been a substantial backlash from many sectors of society, including business and science as well as from politicians. Leading figures, including Sir David Attenborough, wrote to the *Sunday Times* (as noted in Siddique, 2013):

> As the loss of wildlife and habitats continues apace, both in Britain and globally, and as evidence suggests growing numbers of children are missing out

on the mental and physical health benefits of spending time in nature, the place of the natural environment in the national curriculum is more critical than ever. Indeed, the government is committed to nurturing our children's love and respect for nature under two binding international agreements (the United Nations convention on the rights of the child and the convention on biological diversity's Aichi targets). However, under the new draft national curriculum for England, education on the environment would start three years later than at present and all existing references to care and protection would be removed. This is both unfathomable and unacceptable. Today's children are tomorrow's custodians of nature. There is a duty to ensure that all pupils have the chance to learn about threats to the natural world, to be inspired to care for it and to explore ways to preserve and restore it. These proposals not only undermine our children's understanding and love of nature, but ultimately threaten nature itself.

(Nature Vital to Schools, 2013)

The Royal Meteorological Society were equally unequivocal in the response to the consultation: "We strongly advocate that the topic of climate change be added to the geography curriculum to complement coverage of the scientific basis of climate change in the science curriculum" (Ward, 2013).

Substantial numbers of young people also expressed their opposition to the proposed curriculum. Some years before Greta Thunberg rose to prominence, Esha Marwaha, a 15-year-old school student from Hounslow in west London started a petition that was signed by more than 12,000 people. Marwaha wrote that:

Climate change is the most pressing and threatening issue to modern-day society. Through lack of understanding from generations before us, we are having to fix it. And how can we do this without education?

Our government, part of the generation who bear much of the responsibility for this problem, intends to not only fail to act on climate change themselves but to obscure the truth from children and young people. It is outrageous that Michael Gove can even consider the elimination of climate change education for under-14s. We must keep climate change in the curriculum in order for young people take on this challenge of tackling the threat posed by our changing climate.

(Vidal, 2013)

Faced with opposition on a wide range of fronts, it is perhaps not surprising that in July 2013, it was announced that Gove had decided not to drop climate change from the geography curriculum. Writing in the *Guardian*, Patrick Wintour wrote that:

The education secretary's decision represents a victory for Ed Davey, the energy and climate change secretary, who has waged a sustained battle in

Whitehall to ensure the topic's retention. The move to omit it from the new curriculum took on a symbolic status. Gove insisted it was part of his drive to slim an unwieldy curriculum down, to give teachers greater freedom to show their initiative. It was claimed that climate change would appear under science. But environmentalists and science teachers claimed the omission would downgrade the topic and make its existence a matter of greater dispute.

(Wintour, 2013)

The explanation for this change of mind was put down to internal Conservative party issues:

There has been a changing mood in parts of the Conservative party over climate change, with important figures such as Peter Lilley and Lord Lawson challenging its whole premise and the science behind it. Some feared that Gove was putting up a resistance to Davey to pander to the right. But Gove remains a party moderniser and at one point saw the greening of the party as central to its electability.

(Wintour, 2013)

So Bob Ward's concern in 2011 that there were signs of "a political agenda being exercised" turned out to have been a sound prophecy. Some would argue that this story illustrates why control of the curriculum should be taken out of the hands of politicians and be put under the aegis of an independent body.

While this chapter focuses on school education, it is worth noting that some UK universities have introduced climate change degree programmes. Students at John Moores University and Greenwich University could take degrees in Climate Change from 2021 and Northampton University launched its degree in 2022. Many other universities have been offering modules and courses on climate change for several years.

Finally, it is important to note that climate change science has been politicised across the globe. In 2003, it was reported, "White House officials have undermined their own government scientists' research into climate change to play down the impact of global warming" (Harris, 2003).

Emails and internal government documents obtained by The Observer show that officials have sought to edit or remove research warning that the problem is serious. They have enlisted the help of conservative lobby groups funded by the oil industry to attack US government scientists if they produce work seen as accepting too readily that pollution is an issue.

The president at the time, George W. Bush, had once been an oil company executive, as had his vice-president, Dick Cheney. Upon his election to the presidency, Bush withdrew the US from the Kyoto treaty, which had set limits on CO_2 emissions. Some years later, President Trump withdrew the US from the Paris treaty.

The debate continues

Student-led calls for more effective climate change education have increased over the last few years. In 2018, three school students, Anna Taylor, Ivi Hohmann and Daniela Torres Perez, founded the UK Students Climate Network (UKSCN) to lobby the government to take more action on climate change. One strategy they advocated was school strikes, in line with the School Strike for Climate Movement, also known as Fridays for Future, and Youth Strike for Climate, the movement associated with Greta Thunberg. This approach has had a substantial impact, with reportedly more than 1 million students striking in more than 120 countries in March 2019 (Carrington, 2019). Subsequent coordinated actions led to between 4 and 6 million people (depending on whose figures you believe) striking in September of the same year (Taylor et al., 2019). While these strikes have been aimed at forcing governments to listen to younger people's voices, some specific initiatives have focused on changing the curriculum to increase the amount of climate change education taught in schools.

The "Teach the Future" campaign was launched in late 2019 at a Climate Emergency Conference organised by the National Education Union. The initiative is a joint campaign involving the UKSCN and Students Organising for Sustainability (SOS-UK) and it aims to repurpose the education system around the climate emergency and ecological crisis. The organisation's vision of what it wants to see is as follows:

> Students need to be taught about the climate emergency and ecological crisis: how they are caused, what we can do to mitigate them and what our future lives and jobs are going to look like due to them. Sustainability and these crises need to become key content in all subject areas. Educators need to be trained in how to teach about these difficult topics in a way that empowers students, and they need funding and resources to do this.
>
> (Teach the Future, n.d.)

The campaign has three key asks of government:

> ASK 1: A government commissioned review into how the whole of the English formal education system is preparing students for the climate emergency and ecological crisis;
>
> ASK 2: Inclusion of the climate emergency and ecological crisis in teacher training and a new professional teaching qualification;
>
> ASK 3: An English Climate Emergency Education Act.
>
> (Teach the Future, 2020)

In the run-up to the 2019 General Election, the campaign focused on influencing the manifestos of the major political parties and achieved some success, with the Labour Party promising a review of the curriculum. The campaign operates in England, Scotland and Wales.

Teachers' views on climate education

Let us turn now to the views of teachers on climate change and climate change education. In 2019, 86% of teachers surveyed in the US thought that climate change should be taught in school (Ipsos/NPR, 2019). Only 8% of the 505 teachers interviewed did not think that climate change should be taught. The randomly-selected sample included 207 teachers who currently taught climate change and 281 teachers who were not teaching climate change to their students. The top three reasons identified by those teachers who were not currently teaching about climate change were: 1. "It's not related to the subject(s) I teach" (65%); 2. "Students are too young" (20%); 3. "I don't know enough about it" (17%). Just over half the teachers felt that climate change education should start in kindergarten (9%) or elementary (primary) school (44%).

In the same year, a survey of 352 UK teachers, carried out by YouGov for Oxfam and the UK School Climate Network (UKSCN), found that 69% wished to see more teaching in schools about climate change. A total of 70% of respondents supported radical change in UK legislation to make the education system fit for the times in which we live (Oxfam/UK Student Climate Network, 2019). While 89% of the teachers agreed that UK students should always be taught about climate change, its implications for environments and societies around the world and how these implications can be addressed, only 18% said that they had received adequate training to teach about climate change and its impacts.

In a more recent study involving 626 primary and secondary teachers, 73.7% were teaching about climate change or talking to their students about it. This figure is higher than that found in the US survey and may reflect the higher profile of climate change in the UK media including the coverage of the school strikes. Most respondents (both primary and secondary) said that the science, root causes, broader impacts and the issues of social justice associated with climate change should be taught from primary school, which contrasts with how the topic is covered in the national curriculum.

In conclusion

The debate about the position and status of climate change education, which can be traced back to at least the 1990s, continues. The evidence suggests that teachers would support a greater emphasis on climate change education generally and specifically at primary school, where it is almost invisible in the curriculum. Students continue to push for substantially more climate change education, and they are already having some impact on policymakers. The increasing public and political awareness of the impact of climate change on the world means that it is likely that in the near future, climate change education will have a greater presence in the science curriculum in the UK and elsewhere. How this might relate to the role of sustainability education in the curriculum is unclear, but climate change education without sustainability education would seem to be missing an obvious synergy.

Questions for further debate

1 What influence should politicians have on the science curriculum?
2 What should be taught about climate change?

Suggested further reading

Lehtonen, A., Salonen, A. O., & Cantell, H. (2019). Climate change education: A new approach for a world of wicked problems. In *Sustainability, human well-being, and the future of education* (pp. 339–374). Palgrave Macmillan.

Monroe, M. C., Plate, R. R., Oxarart, A., Bowers, A., & Chaves, W. A. (2019). Identifying effective climate change education strategies: A systematic review of the research. *Environmental Education Research, 25*(6), 791–812.

Stevenson, R. B., Nicholls, J., & Whitehouse, H. (2017). What is climate change education? *Curriculum Perspectives, 37*(1), 67–71.

Verlie, B. (2020). From action to intra-action? Agency, identity and 'goals' in a relational approach to climate change education. *Environmental Education Research, 26*(9–10), 1266–1280.

References

Bell, A. (2013, March 19). Yes, you should be worried about climate change education. *The Guardian.* https://www.theguardian.com/science/political-science/2013/mar/19/science-policy

Boffey, D. (2012, June 12). Michael Gove's own experts revolt over 'punitive' model for curriculum. *The Guardian.* https://www.theguardian.com/politics/2012/jun/17/michael-gove-national-curriculum

Carrington, D. (2019, March 19). School climate strikes: 1.4 million people took part, say campaigners. *The Guardian.* https://www.theguardian.com/environment/2019/mar/19/school-climate-strikes-more-than-1-million-took-part-say-campaigners-greta-thunberg

Harris, P. (2003, September 21). Bush covers up climate research. *The Guardian.* https://www.theguardian.com/environment/2003/sep/21/usnews.georgewbush

Hickman, L. (2013, March 18). Teachers should be given free rein to teach climate change in schools. *The Guardian.* https://www.theguardian.com/environment/blog/2013/mar/18/teachers-climate-change-schools-curriculum

IPSOS/NPR. (2019, April 22). *Teachers agree that climate change is real and should be taught in schools* [Press release]. https://www.ipsos.com/sites/default/files/ct/news/documents/2019-04/npr_teachers_climate_change_topline_april_22_2019.pdf

Jowitt, J. (2013, March 17). Climate debate cut from national curriculum for children up to 14. *The Guardian.* https://www.theguardian.com/environment/2013/mar/17/climate-change-cut-national-curriculum

Lester, S. (2013, March 19). Why it's a mistake to trim climate change from the curriculum. *The Independent.* https://www.independent.co.uk/voices/comment/why-it-s-a-mistake-to-trim-climate-change-from-the-curriculum-8540423.html

Nature Vital to Schools. (2013, April 14). *The Sunday Times.* https://www.thetimes.co.uk/article/letters-and-emails-april-14-9djzx08kxmm

Oxfam/UK Student Climate Network (UKSCN). (2019). *Climate change education.* https://bit.ly/3uK0mxw

Scott, W. A. H. (2011, June 13). Studying climate's ok (probably); but studying climate change definitely isn't. *Bill Scott's Blog.* https://blogs.bath.ac.uk/edswahs/2011/06/13/studying-climates-ok-probably-but-studying-climate-change-definitely-isnt/

Shepherd, J. (2011, June 12). Climate change should be excluded from curriculum, says adviser. *The Guardian.* https://www.theguardian.com/education/2011/jun/12/climate-change-curriculum-government-adviser

Siddique, H. (2013, April 14). Plans to drop climate debate from national curriculum 'unacceptable'. *The Guardian.* https://www.theguardian.com/environment/2013/apr/14/plans-drop-climate-change-curriculum

Slingo, J. (2020, September 4). Portuguese children sue 33 countries over climate 'failures'. *The Law Society Gazette.* https://www.theguardian.com/law/2020/sep/03/portuguese-children-sue-33-countries-over-climate-change-at-european-court

Smith, J. (2013, March 5). Less climate change: More flag quizzes. *Citizen Joe Smith.* https://citizenjoesmith.wordpress.com/2013/03/05/less-climate-change-more-flag-quizzes/

Taylor, M., Watts, J., & Bartlett, J. (2019, September 27). Climate crisis: 6 million people join latest wave of global protests. *The Guardian.* https://www.theguardian.com/environment/2019/sep/27/climate-crisis-6-million-people-join-latest-wave-of-worldwide-protests

Teach the Future. (2020). *Teach the future.* https://uploads-ssl.webflow.com/5f880 5cef8a604de754618bb/5fa3e667cd1ee8abe322f067_Asks%20(England).pdf

Teach the Future. (n.d.). *Teach students about climate change.* https://www.teachthefuture.uk

Vidal, J. (2013, March 21). Thousands sign school climate change petition started by 15-year-old. *The Guardian.* https://www.theguardian.com/environment/2013/mar/21/school-climate-change-petition

Wals, A. E. J., Brody, M., Dillon, J., & Stevenson, R. B. (2014). Convergence between science and environmental education. *Science, 344*, 583–584.

Ward, B. (2013, April 29). PR smokescreen cannot hide the holes in climate teaching proposals. *The Guardian.* https://www.theguardian.com/environment/blog/2013/apr/29/pr-smokescreen-holes-climate-teaching-proposals

We Need Stretching Exams That Compare with the World's Most Rigorous. (2011, June 18). *The Times.* https://www.thetimes.co.uk/article/we-need-stretching-exams-that-compare-with-the-worlds-most-rigorous-cbcxzpvbv33

Wintour, P. (2013, July 5). Michael Gove abandons plans to drop climate change from curriculum. *The Guardian.* https://www.theguardian.com/education/2013/jul/05/michael-gove-climate-change-geography-curriculum

Chapter 6

Science education for citizenship

Contributions from knowledge *of* and *about* science in the context of the COVID-19 pandemic

Rosária Justi, Poliana Maia and Monique Santos

Citizenship in the 21st century

Before we start discussing issues related to science education for citizenship, and as a way to support such a discussion, the meaning we assume for *citizenship* should be made clear. In any dictionary, the general meaning attributed to citizenship is related to "the status of being a member of a particular community, resulting in enjoying rights and assuming duties". According to the *Stanford Encyclopedia of Philosophy*, although this broad definition has been used since the 18th century, the concept has been controversial, which is associated with the three main elements or dimensions that compose it: "citizenship as legal status, defined by civil, political and social rights", the consideration of citizens "as political agents, actively participating in a society's political institutions", and "citizenship as membership in a political community that furnishes a distinct source of identity" (Leydet, 2017, p. 2). However, it is also recognised that the legal status has been the main one, which implies citizenship meaning membership in a community. Unfortunately, in the current world, having the right to vote does not mean that all citizens from a given democratic society have the same conditions to participate in community decisions or to equally enjoy the same social rights.

On the other hand, assuming that society is something mutable, that we build each community that comprises it on a daily basis, and that in these first decades of the 21st century we should certainly think as a global society, the United Nations Educational, Scientific and Cultural Organization (UNESCO) has recently published a series of books discussing the topic *global citizenship education* (GCE) (UNESCO, 2014, 2015, 2018). For UNESCO:

> [t]he notion of "citizenship" has been broadened as a multiple-perspective concept. It is linked with growing interdependency and interconnectedness between countries in economic, cultural and social areas, through increased international trade, migration, communication, etc. It is also linked with our

DOI: 10.4324/9781003137894-7

concerns for global well-being beyond national boundaries, and on the basis of the understanding that global well-being also influences national and local well-being.

(UNESCO, 2014, p. 14)

This perspective turns the focus from the legal status to:

a sense of belonging to a broader community and common humanity, promoting a "global gaze" that links the local to the global and the national to the international. It is also a way of understanding, acting and relating oneself to others and the environment in space and in time, based on universal values, through respect for diversity and pluralism.

(UNESCO, 2014, p. 14)

So, citizens' active participation in society involves social responsibility. Thus, it can be viewed as intrinsic to the life of human beings and going far beyond legal rights and duties.

Education and the promotion of citizenship

As part of its efforts to contribute to the implementation of global citizen education (GCE) across countries, and acknowledging "the civic, social, and political socialization function of education" (UNESCO, 2014, p. 15), as well as its role as a transformative agent in society, UNESCO emphasises that "knowledge, skills and values for the participation of citizens in, and their contribution to, dimensions of societal development which are linked at local and global levels" (UNESCO, 2014, p. 15) should be the focus of GCE. Thus, GCE entails cognitive, socioemotional, and behavioural conceptual dimensions, which characterise it as a complex and challenging initiative to educate people to face the challenges of our interconnected world. Also, being consistent with a critical democratic discourse, GCE has among its aims:

- develop and apply critical skills for civic literacy, e.g., critical inquiry, information technology, media literacy, critical thinking, decision-making, problem solving, negotiation, peace building and personal and social responsibility;
- recognise and examine beliefs and values and how they influence political and social decision-making, perceptions about social justice and civic engagement;
- develop values of fairness and social justice, and skills to critically analyse inequalities based on gender, socio-economic status, culture, religion, age and other issues;
- participate in, and contribute to, contemporary global issues at local, national and global levels as informed, engaged, responsible and responsive global citizens.

(UNESCO, 2015, p. 16)

On the other hand, some people may view GCE as a kind of utopian perspective in a world where, by valuing individual achievements and self-investment, neoliberalism exerts a great influence in many sectors, including the educational one (Pais & Costa, 2017). From the neoliberal view, criticising, questioning, and generating proposals for a better world are not among the aims of individuals' education. Rather, education would be guided by values of individual success such as competitiveness, individualism, and self-investment.

We mention these two perspectives to claim that our ideas are aligned with the critical democratic discourse, but that we also recognise that GCE should play the role of promoting the recognition of, and criticisms on, the neoliberal values imposed and widely disseminated in many countries that reinforce social inequalities.

What has science education to do with citizenship?

Nowadays, even when we do not realise it, science and technology pervade our lives and influence many of our decisions. In the current pandemic era, the relevance of science for the promotion of human well-being seems unquestionable for most of those who read newspapers or interact with any other media. For those involved with science education on any level, it has been common to read forceful and vehement defences of the roles it can play in the education of 21st-century citizens. For instance, a report of the European Commission, produced by a group of experts on science education, synthesises such roles, stating that:

> Science education is vital: [t]o promote a culture of scientific thinking and inspire young people in using evidence-based reasoning for decision-making, as opposed to values and reasoning processes that are less reliable or that are only based on beliefs or feelings; [t]o ensure citizens have the confidence, knowledge and skills to participate actively in an increasingly complex scientific and technological world; [t]o develop the competencies for problem-solving and innovation, as well as analytical and critical thinking that are necessary to empower citizens to lead personally fulfilling, socially responsible and professionally-engaged lives promoting solidarity at national, European and global level; . . . [t]o empower active and responsible participation in public science communication, debates and decision-making as active engagement of European citizens in the big challenges facing humanity today.
> (Hazelkorn et al., 2015, pp. 14–15)

But how may these aims – which are closely related to those of the GCE – be reached? How may science teachers promote citizenship in their classes?

In principle, we identify three fronts from which those aims could be addressed in science teaching: *scientific content itself*, *aspects of nature of science*, and *science teaching methodologies*. In terms of science content, it seems clear that the acquisition of static disciplinary knowledge presented as ready truths should not be prioritised (Talanquer et al., 2020) as it does not support the development of skills required of 21st-century citizens. Thus, the outcome of teaching focused on traditional science content coverage is not a well-informed citizen. Content should certainly include the core ideas of the distinct scientific disciplines, but those core ideas should be approached from an interdisciplinary perspective and connected to scientific practices involved in their generation and/or use. This would contribute to contextualise the content, to justify its learning in a way that allows students to understand "why we believe what we know in contrast to alternatives" (Duschl & Grandy, 2013, p. 2114), and to support deep understanding, critical analysis, reflections, and generation of relevant questions concerning a given idea, event, or phenomenon.

In the same vein, knowledge of nature of science (briefly, of how science works and interrelates to society) cannot be restricted to declarative tenets about scientific knowledge. Many researchers (for instance, Allchin, 2020; Duschl & Grandy, 2013; Hodson, 2014; Kolstø 2001; Nielsen, 2013) have emphasised that the inclusion of discussions about how scientific knowledge is produced, validated, communicated, and used contribute to enhance the education of creative and responsible citizens.

Finally, the way knowledge *of* and *about* science is taught may be essential for attaining GCE. Activities experienced by science students should require them to be creative and independent, to ask relevant questions, to evaluate and relate data and other types of information, to understand the importance of, and how to develop, a *scientific attitude*; that is, to be committed with both caring about any kind of evidence (e.g., historical, qualitative, statistical) and changing ideas in light of new and robust evidence (McIntyre, 2019). From participating in activities from such perspectives, students may learn to think critically and to plan responsible actions in socioscientific contexts. Additionally, they should foster students' engagement in explicit and contextualised reflections on issues *about* science and its role in society. However, being science teachers, all of us know that teaching methodologies depend not only on teaching aims for each students' age (that may be shared among communities of teachers), but mainly on teaching contexts (especially characteristics of the students). Therefore, even though we view teaching methodology as a highly potential contributor to citizenship education, we leave to each teacher the decisions on methodological aspects for teaching the issues discussed in this chapter. As for such issues, in the second part of the chapter, we detail the relationships between the teaching *of* and *about* science and citizenship. In so doing, we build a case concerning one aspect of the current pandemic and use several of its elements as examples to illustrate our arguments.

Teaching *of* and *about* science and citizenship

The case of COVID-19 vaccines

On 11 March 2020, the World Health Organization (WHO) classified the COVID-19 outbreak as a pandemic. Since then, many resources and efforts have been utilised by the WHO, governments, and private institutions, from individual countries or working together, aimed at stemming the pandemic, bringing down the mortality rate, and reducing the effects of the virus (including the economic impact that has been triggered worldwide). In this sense, for example, there have been studies related to the production and use of protective equipment, forms of disease contagion, treatments for patients, production of vaccines, and the effects of physical isolation.[1] Even though all of them are important topics to be discussed, because mass vaccination is the only way to end the pandemic, here we focus on vaccines to support a more detailed discussion about the relationships between the teaching *of* and *about* science and citizenship.

Generally speaking, vaccines can be defined as substances that prevent people or animals from getting a disease by training their immune system to recognise pathogens (a virus or a bacteria) that may cause diseases. Then, the body fights the infection without getting sick. Since the development of the first vaccine in the 18th century, there have been anti-vaccination (or anti-vaxxing) movements, whose origins (many of them motivated by religious beliefs) included lack of understanding about the meaning of a vaccine or how different types of vaccines work, as well as attempts to minimise the effects of the disease or to convince people that the vaccines were more dangerous than the disease itself. However, in the last decades, it is clear that such movements have been largely disseminated by fake news and misinformation, which are easily posted, visualised, and shared by anyone thanks to the wide reach of contemporary media through the Internet. In particular, social media platforms, sometimes supported by 'celebrities' or even physicians (who are always introduced as great experts on a particular subject), have been largely used to spread fake news. Their main focuses continue to be unclear harms caused by vaccines and appeals to how vaccination infringes personal, religious, or social rights. However, they do not present data that supports the anti-vaccination position. In the current context, an example of fake news is detractors' claims that vaccines using messenger RNA to stimulate immune response can multiply out of control and alter a person's DNA. As for misinformation, a common example is the claim that people should not trust the COVID-19 vaccines because they have been developed too fast.

It is true that one year after the WHO declared COVID-19 a pandemic, there are six vaccines approved for full use and six authorised for emergency use in different parts of the world. Usually, the development of a vaccine involves years of research and testing before attaining any kind of approval (Zimmer et al., 2021). However, this does not mean those vaccines, as well as more than a hundred others that are undergoing different stages of trials, cannot protect people because they were produced in a shorter time. It also does not mean that the evaluations of their efficacy and safety are being compromised.

In the case of the COVID-19 vaccines, their development and approval were made possible due to three main factors. First, none of them had to be developed from scratch. Conversely, all of them were based on previous knowledge of vaccines against other viruses, or on studies aimed at producing vaccines for other diseases (mainly those caused by other coronaviruses, like SARS-CoV and MERS). This is true even for those based on newer technologies, such as the use of nucleic acids. Second, scientists from a number of areas, from different universities, institutions, and pharmaceutical companies, in the same or different countries, have been working together (sometimes coordinated by global organisations, like the WHO), sharing empirical results and knowledge produced. Such collaboration has also been possible from the decision of all major publishers and scientific organizations to focus their efforts on bringing out wide dissemination of research on the subject. In this sense, there have been: specific calls for papers about COVID-19 in journals from several areas; exemption from fees for submission of papers related to COVID-19 in journals that have fees for submissions; and the release of free access to publications related to the pandemic. Third, the scientists' works have been receiving all the necessary financial support – at least in countries where science is being given top priority now.

The crucial aim of all the efforts has been the development of vaccines that have better immune responses and maintain the immune response for a longer period. But why there are so many types of vaccines? How do they differ? Basically, there are different types of vaccines because they are produced from distinct technologies. Until recently, the most common types of vaccines were *live attenuated* ones, produced from a weakened form of the disease-causing organism, or *inactivated* vaccines, which use a destroyed genetic material of the virus that cannot replicate (e.g., vaccines from the Chinese companies Sinovac Biotech and Sinopharm). In both cases, these forms of the pathogen trigger an immune response. Both are well-established technologies. Similar technology is used in the production of the *subunit* or *acellular* vaccines, which contain purified pieces of the pathogen (often fragments of one of its proteins) that are able to stimulate immune cells (as in the Novavax (USA)). Newer technologies (that have been studied for decades but were licenced for human application for the first time now) use the body's own cells to produce antigens that trigger the response when detected by the immune system. The outcomes are (i) *nucleic acid* vaccines, which contain genetic material from the pathogen that instructs cells to produce antigens (for example, the successful vaccines from Pfizer-BioNTech [USA and Germany] and Moderna [USA]); and (ii) *non-replicating viral vector* vaccines, in which another vector (in the case of COVID-19, the adenovirus that causes the common cold) gives such instructions (vaccines from Oxford-AstraZeneca [UK], Gamaleya [Russia], and Jansen-Cilag [a Belgian subsidiary of Johnson & Johnson]). The efficacy rate of those vaccines is different (varying from 50 to 95%), but the essential points now are to slow the spreading of the virus and to reduce the severity of the disease – to which all of them can contribute.

Before a vaccine gets approval, it must undergo a long process. After the choice of the technology and the production of the initial raw material, it is tested on cells and given to animals (mainly monkeys, who are phylogenetically close to humans) so that its immune response can be analysed (the preclinical stage). After the preclinical stage, if the vaccine proves to be safe and promotes a good immune response, it goes on to the three-stage clinical trial process. In phase 1, the vaccine is inoculated in a small group of people to analyse its safety and dosage, and to confirm that it does indeed stimulate the immune system. Then, in phase 2, hundreds of people from distinct age groups receive the vaccine in small doses and are monitored so that potential different reactions can be identified. Based on the results, scientists can determine the dosage that would be necessary to produce the desired response. Finally, phase 3 is double-blind and randomised. Thousands of volunteers from several locations participate in this phase, some receiving the actual vaccine while others receive a placebo. Then, scientists wait some time to identify how many people from each group become infected, and if there were any relatively rare side effects. At the end of this phase, the clinical trial results are analysed by the regulators (a multidisciplinary team of scientists) in each country to decide, based on desired statistical results, whether the vaccine is going to be approved or not. This does not mean the end of the process, as safety continues to be monitored during the vaccination of the population.

Vaccination is another stage of the process. Besides the difficulties in producing a huge number of doses in the shortest possible time and the complexity of the logistical challenges to distribute and store the vaccines, the population has to be convinced to be vaccinated. Unfortunately, in the current post-truth era, this may be the most challenging step in some contexts. To persuade the population to be vaccinated, the results concerning safety and efficacy rate, as well as other information about how each vaccine works, how it should be administrated, any possible adverse events, etc., have to be clearly communicated. Most people have difficulties understanding such information as they are not trained in science; they have neither experienced scientific practices, nor thought about how science and technologies are produced and how they reach them. Additionally, many people remain skeptical about everything related to science, mainly when it involves their health and their previous beliefs (especially when they echo claims disseminated through social media). Therefore, experts (scientists themselves) and non-experts (journalists and science communicators) play a key role in providing accurate and reliable information in a trustworthy and accessible way to the general public, in order to avoid (or at least minimise) gaps in the communication, as well as the emergence and/or nurturing of fake news.

Relevant issues to be discussed

Much more could be said about the COVID-19 vaccines, and certainly many of our readers have further relevant information about it. Our option for presenting this brief case is aimed at organising some selected information that may help us to illustrate and support our arguments.

All of us, science teachers and researchers, know from both teaching experiences and empirical data that many students have difficulties learning science content. A large number of reasons may cause such difficulties; common ones including the presentation of the content as unquestionable truths in the form of concepts and relationships among them (sometimes represented in (complex) equations, graphs, and/or other forms of representation), and the non-insertion of the content in any context that could be significant to students. Many of us have already taught concepts like virus, antigen, genetic material, or the process of how viruses act in our body or how our immunologic system works, and we realised that students struggle to make them meaningful. In other words, it is difficult for students to integrate them into their knowledge not in a declarative way, but rather being able to apply them to understand contemporary contexts, to analyse events or news involving them, and to position themselves when faced with such information.

Even our brief presentation of the case of the COVID-19 vaccines makes it evident that those who do not build their knowledge of these and other related concepts and processes are likely to either not be able to understand crucial issues concerning the pandemic situation and be aware of the risks it may bring to us, or to be easily influenced by fake news. One of the common approaches to producing fake news is the use of scientific terms/ideas whose meanings are not understood by ordinary people, who tend to believe them, thinking they are supported by 'the authority of science'. Someone who just believes in ideas supposedly related to science without understanding their acceptable scientific meaning or how they were developed and became acceptable would certainly not think about questioning them, or consider how to develop and apply most of the critical skills identified by UNESCO as essential to become a global citizen. Therefore, the first step to being taken by science teachers committed to foster students' citizenship education is to approach the core concepts and relationships in a way that makes them meaningful and relevant to students.

In addition to what has been stated, a series of aspects *about* science can emerge from a broad discussion of specific points of the case of the COVID-19 vaccines having the potential to support citizenship education. For instance, crucial points emphasised in the case are the development of distinct vaccines in less than a year. As previously mentioned, among many other reasons, the new knowledge and technologies that made these processes possible were developed from previous ones, thus showing that scientific knowledge is not constructed in one go. Conversely, its development is an ongoing and non-linear process, in which each scientist has to be motivated, creative, and intelligent, and that groups of scientists (from different areas) must interact with each other, sharing information, ideas, doubts, uncertainties, and mistakes. In so doing, they are able to define the specific logic of reasoning and routes of actions (which may involve distinct doses of rationality and subjectivity in specific moments and steps of the process). This set of aspects shows that the development of distinct vaccines is a process comprised of several epistemic practices based on trustworthy evidence. It also shows that scientific knowledge is tentative and complex – which highlights the crucial importance of validation processes (that have been conducted with extreme rigour), ensuring such knowledge can be accepted. These aspects go beyond only methodological procedures to produce knowledge and technologies. By being aware of all of them, students will be well-informed about this context and able to develop and apply both critical inquiry (for instance, by checking conflicting information when seeking diverse and reliable sources of information) and critical thinking (which requires people to question reasons and justifications that support what they believe in (Vieira et al., 2011)). Both skills can help them to position themselves in related discussions.

Also, the whole process of development of the COVID-19 vaccines would not be possible without the existence of funding sources from diverse origins that are investing a lot of money and providing top lab structures. These economic aspects have been essential in all phases of the process (from initial production to clinical trials and mass vaccination) so that each of them could have occurred in the shortest possible amount of time, but performed with the necessary rigour and ethics to produce reliable outcomes for their desired goals.

Becoming aware of the level of influence economic aspects have on the development of vaccines may be an additional element to support students' critical thinking and to contribute to their recognition of the value of such influences. On the other hand, economic aspects involved, mainly in the distribution of vaccines, clearly show inequalities between countries from different continents, or even from the same continent. So becoming aware of that other side of the coin may play a key role in the development of students' values of fairness and social justice, as also claimed by UNESCO as an important aim of GCE.

Another crucial issue involved concerns the ethics, especially related to the behaviour of scientists, above all in strictly following the rules of academic production and publication, and volunteers, by reporting true information when participating in the clinical trials. Ethical aspects also involve the commitment of different people in reporting true information to distinct publics, mainly in communications from scientists, physicists, members of governments or other relevant institutions, journalists, or other science communicators. Regarding communication among scientists or from them to the general public, care should be taken with what information is released, particularly in terms of its accuracy and reliability.

Beyond these ethical issues, there are the socio-political ones, including not only the importance of reliable information targeting the general public, but also how information is communicated – which seems essential when addressing the big challenge of preparing people to deal with so much information, including fake news. This point has been particularly relevant because fake news is generally made through emotional appeals and employing powerful imagery or other forms of representation, trying to reach two main groups of people: those who do not have the knowledge or skills to critically analyse and question such information, and vulnerable people subject to ideologic isolation; that is, those who access only news that reinforces their own previous ideas (Puri et al., 2020). That is another source of concern, because such reinforcement can also occur due to the Internet's own search algorithms, which guide the content of what people access based on past access patterns. In a socio-political bias, the intense participation of people in spreading news – unfortunately sometimes fake information – can compromise the potential of mass vaccination to change peoples' current way of life. So, raising students' awareness of such issues may favour their media and information literacy, meaning to instrumentalise them to distinguish reliable information from fake news, to identify where and how to get relevant information, to critically evaluate data and modes of representation of information, to recognise bias and embedded values in given media and information, and to use them in an ethical and socially responsible way.

The case of the COVID-19 vaccines also highlights another essential aspect of science: its non-neutrality. Science influences and is influenced by a series of factors, like individual or groups of scientists' motivation to solve a problem and the people's willingness to be volunteers in clinical trials of vaccines, even knowing the risk involved. This last point can be analysed through the perception of distinct cultural views as, in some countries, people are used to being volunteers, including being involved in tests of medicines, as a way to participate in science development.

Other factors that can determine the way science can impact on society are the political ones. This issue can be clearly exemplified by official agencies' role in the approval of vaccines, as well as the distinct levels of governments' incentives in their production and actions to provide a quick vaccination of the population.

All these aspects make the interrelationships of science and society evident. Therefore, their introduction in science lessons seems essential, and in some sense natural, when teachers are concerned about promoting students' responsible and participatory citizenship. On the other hand, we are not advocating that teachers should plan activities to discuss all these aspects with students from all educational levels, or at only one moment. To reach science education aims and/or those concerning the promotion of citizenship, teachers may use their knowledge and previous experiences to identify which aspects are worth being discussed in specific contexts with a given group of students. In line with those aims, we argue in favour of a contextualised, explicit, and integrated (with the discussion of content knowledge) manner of introducing aspects *about* science in science lessons (Santos et al., 2020).

Concluding thoughts

Our main aim when writing this chapter was to support teachers' reflections about citizenship education and the role of science education in promoting it. We acknowledge this may be addressed from three fronts: approaching scientific content from interdisciplinary and contextualised perspectives, highlighting a diversity of aspects concerning nature of science, and using teaching methodologies that engage students in participatory and open-minded learning processes. To make our arguments clear, we built a brief case about what is certainly one of the landmarks of the 21st century: the development and application of the COVID-19 vaccines. Then, we used its elements as evidence of our claims in favour of a comprehensive science education. From such an education, students may be motivated not only to develop "positive but critical attitudes towards science" (Hodson, 2014, p. 945), but also to learn and enact citizenship.

Hoping that we have been successful in providing input for teachers' meaningful reflections, as a final remark, we invite teachers to actively engage in this debate and in promoting science education from this perspective having in mind that the education of well-informed, creative, critical, personally and socially responsible, and actively participating global citizens is not a utopia. Rather, it may certainly be our main contribution to building a healthy and peaceful world for our and future generations.

Questions for further debate

1 What role should science education have in citizenship education?
2 How could science teachers use events like the COVID-19 pandemic to teach about citizenship?

Suggested further reading

Valladares, L. (2021). Scientific literacy and social transformation. *Science & Education, 30*(3), 557–587. https://doi.org/10.1007/s11191-021-00205-2

Yacoubian, H. A., & Hansson, L. (2020). *Nature of Science for Social Justice.* Springer.

Note

1 Although the phrases 'social distancing' and 'social isolation' are commonly used in the academic literature and by the media, as El-Hani and Machado (2020) do, we chose to use physical distancing or isolation. Following the Mind the Gap campaign, we recognise the importance of developing a more sensitive look at those who in fact are in social distance, or isolation, such as people with psychosocial and/or physical disabilities, elderly people who are in nursing homes, and inmates. In addition, social media has provided the approximation of people, despite the physical distancing or isolation.

References

Allchin, D. (2020). From nature of science to social justice: The political power of epistemic lessons. In H. A. Yacoubian & L. Hansson (Eds.), *Nature of science for social justice* (pp. 23–39). Springer. https://doi.org/10.1007/978-3-030-47260-3_2

Duschl, R. A., & Grandy, R. (2013). Two views about explicitly teaching nature of science. *Science & Education, 22*(9), 2109–2139. https://doi.org/10.1007/s11191-012-9539-4

El-Hani, C. N., & Machado, V. (2020). COVID-19: The need of an integrated and critical view. *Ethnobiology and Conservation, 9*(18), 1–20. https://doi.org/10.15451/ec2020-05-9.18-1-20

Hazelkorn, E., Ryan, C., Beernaert, Y., Constatinou, C. P., Deca, L. Grangeat, M., Karikorpi, M., Lazoudis, A., Casulleras, R. P., & Welzel-Breuer, M. (2015). *Science education for responsible citizenship*. European Commission. https://doi.org/10.2777/12626

Hodson, D. (2014). Nature of science in the science curriculum: Origin, development, implications and shifting emphases. In M. R. Matthews (Ed.), *International handbook of research in history, philosophy and science teaching* (pp. 911–970). Springer. https://doi.org/10.1007/978-94-007-7654-8_28

Kolstø, S. D. (2001). Scientific literacy for citizenship: Tools for dealing with the science dimension of controversial socioscientific issues. *Science Education, 85*(3), 291–310. https://doi.org/10.1002/sce.1011

Leydet, D. (2017). Citizenship. In E. N. Zalta (Ed.), *The Stanford encyclopedia of philosophy* (Fall 2017 Edition). https://plato.stanford.edu/archives/fall2017/entries/citizenship/

McIntyre, L. (2019). *The scientific attitude*. MIT Press.

Nielsen, K. H. (2013). Scientific communication and the nature of science. *Science & Education, 22*(9), 2067–2086. https://doi.org/10.1007/s11191-012-9475-3

Pais, A., & Costa, M. (2017). An ideology critique of global citizenship education. *Critical Studies in Education, 61*(1), 1–16. https://doi.org/10.1080/17508487.2017.1318772

Puri, N., Coomes, E. A., Haghbayan, H., & Gunaratne, K. (2020) Social media and vaccine hesitancy: New updates for the era of COVID-19 and globalized infectious diseases. *Human Vaccines & Immunotherapeutics, 16*(11), 2586–2593. https://doi.org/10.1080/21645515.2020.1780846

Santos, M., Maia, P. F., & Justi, R. (2020). A model of science to base the introduction of aspects of nature of science in teaching contexts and to analyse such contexts. *Revista Brasileira de Pesquisa em Educação em Ciências, 20*, 617–651. http://doi.org/10.28976/1984-2686rbpec2020u617651

Talanquer, V., Bucat, R., Tasker, R., & Mahaffy, P. G. (2020). Lessons from a pandemic: Educating for complexity, change, uncertainty, vulnerability, and resilience. *Journal of Chemical Education, 97*(9), 2696–2700. https://dx.doi.org/10.1021/acs.jchemed.0c00627

UNESCO. (2014). *Global citizenship education: Preparing learners for the challenge of the twenty-first century*. UNESCO. https://unesdoc.unesco.org/ark:/48223/pf0000227729?posInSet=11&queryId=01626846-b4e3-4003-92b6-a0fa4ba1b01b

UNESCO. (2015). *Global citizenship education: Topics and learning objectives*. UNESCO. https://unesdoc.unesco.org/ark:/48223/pf0000232993

UNESCO. (2018). *Preparing teachers for global citizenship education: A template*. UNESCO. https://unesdoc.unesco.org/ark:/48223/pf0000265452?posInSet=13&queryId=01626846-b4e3-4003-92b6-a0fa4ba1b01b

Vieira, R. M., Tenreiro-Vieira, C., & Martins, I. P. (2011). Critical thinking: Conceptual clarification and its importance in science education. *Science Education International, 22*(1), 43–54.

Zimmer, C., Corum, J., & Wee, S.-L. (2021, March 11). Coronavirus vaccine tracker. *The New York Times*. https://nyti.ms/2MHNdRL

Chapter 7

Science and anti-science

Mike Watts

> Science is built up of facts as a house is of stones, but a collection of facts
> is no more a science than a pile of stones is a house.
>
> Henri Poincare, *La Science et l'Hypothese* (1908)

The debate

In this chapter, I debate the following statement: 'It is the role of all science educators to root out and eradicate all science ignorance'. Using expressions such as 'root out' and 'eradicate' is not to be mischievous, but to highlight just how contentious is the interplay between science and 'anti-science'. In exploring these issues, I am rather more interested in the nature of 'science ignorance' than the possibilities of wholesale eradication (either in the short- or long-term) – or who exactly has the responsibility for doing the eradication.

John Dewey set the stage for our current debate in 1910, when he argued that the proper goal of science education was to create 'scientific habits of mind' in all people (Dewey, 1922). Perhaps *all* people is an ambitious programme, and the debate here surrounds a particularly 'difficult' group: those who reject such Deweyesque habits of mind; for example, 'Flat Earthers' and 'Anti-Vaxxers'. What ails them? What 'science educatively' is to be done with them? They are not people traditionally described as having learning difficulties; in fact, they are often extraordinarily adroit and certainly voluble. They do, though, have 'science difficulties'.

We are all on the spectrum

Let me speculate first on a short spectrum, one that runs from '*science refusal*', '*science hesitancy*', '*science responsive*' through to '*science habitué*'. In this chapter, I am interested in the first of these, in people who will have no truck whatsoever with science – at least not in ways they (and we) know it. In large part, I am discussing adults rather than children but, since all adults were once children, their approach to science most probably has roots in childhood and adolescent experiences.

DOI: 10.4324/9781003137894-8

I describe myself as a science habitué in that I live and work within the world of science, I share most though not all of its languages, precepts, mores and customs, and while I position myself as suitably critical, my demeanour is generally very 'science friendly' peppered with 'science wonder' and a degree of 'science awe'. My wife and daughter, on the other hand, have backgrounds in the arts, are much more hesitant about it all, know it has value but have little direct interest and say science is simply 'not for them'. They have, they say, other (better) things to do with their days. Over time and a shared life, I like to think they have moved to a greater appreciation of – and responsiveness to – science and the ways in which it works but, nevertheless, they hesitate well short of actual engagement.

My definition of a 'science refuser', or anti-scientist, is a person who dismisses well established science methods and results for reasons that are not themselves scientifically grounded. I make a distinction between science refusal and true scepticism – the latter challenges science and can prompt revisions, but on the basis of strong evidence and reasoned theorising. After all, true scepticism is at the core of scientific methods. In my category of science refusers, I include people who, for example, maintain that the Earth is flat, but add in those for whom the National Aeronautics and Space Administration (NASA) covertly fabricated the Moon landings in the early 1970s; who think MMR vaccinations cause autism and are a 'Big Pharma' conspiracy; say global warming is a hoax and does not exist; claim that the Covid-19 pandemic is fake news perpetrated to advance a demonic political agenda to microchip the entire population; insist corn circles are caused by aliens who abduct and brainwash people; and so on.

The dilemmas and debates at the heart of this chapter, then, relate to such examples of anti-science. Yes, of course, flat-Earth people have the right to think just what they want, and even – in most situations – to give expression to those thoughts, even if those thoughts are virulently anti-science. Believing that the Earth is flat is relatively harmless. However, people do not (in my view) have the right to endanger the herd by reducing immunisation or by increasing their carbon footprint. They are not (in my view) free to be profligate with scarce natural resources and wasteful of the shared environment. For me, it is not just a case of drawing lines between individualism and collectivism, between the rights of the person and responsibilities towards broader society. After all, I am a science educator, and I am driven to educate people in and about science. So, come along then, dear reader, we have an educational job to do.

One example: arguments for a flat Earth

My reading here leads me to a rather general synthesis of the arguments for a flat Earth – although quite possibly no flat-Earther actually believes all of this. Having been the established assumption since antiquity, the notion of a flat Earth resurfaced in the 1800s as a backlash to scientific progress, principally among those who wished to return to traditional – often religious – teachings. Perhaps the most famous proponent was the British writer Samuel Rowbotham (1816–84). The first argument is that, walking around on the planet's surface, it *looks* and *feels* flat. Our everyday experience is of a flat horizontal surface, and everyday experience is both a blessing and a curse in discussions of scientific explanations. From this initial premise, a case is then built for the Earth being a flat disc with the Arctic Circle in the centre and Antarctica, a 150-foot-tall wall of ice, around the rim. It is said that employees of NASA are used to guard this ice wall to prevent people from climbing over and falling off the disc. Earth's diurnal cycle of light and dark (day and night) is explained by arguing that the Sun and Moon are spheres measuring 50 kilometres each that move in circles 5,000 kilometres above the plane of the Earth. Like spotlights, these celestial spheres illuminate different portions of the planet in a 24-hour cycle. Flat-Earthers believe there must also be an invisible 'anti-Moon' that obscures the Moon during lunar eclipses. What happens below the Earth's surface is described simply as 'rocks'. And gravity is just an illusion; it is merely a natural occurrence that things fall to the ground.

My intention in picking on flat-Earthers is not to present science explanations to refute these points, but to note them and return later in the discussion to consider the 'what and why' of this kind of anti-science. Of immediate interest, though, is the depth and detail of science I would need to draw upon should I choose to challenge these ideas from the relative comfort of my home or classroom. That is, without the prospects of travelling in an aircraft with said flat-Earther to Antarctica and then beyond. Straightforward evidence such as photographs of a spherical planet from space, for instance, are commonly dismissed as a complex NASA conspiracy designed to delude the world's population. The difficulty here is to challenge contrary beliefs, a mistrust of science, particular political perspectives and persuasions, fear, anxiety and other emotions. American writer Mark Twain lamented the fact that people's beliefs and convictions are usually secondhand, taken without examination from others who have not themselves examined the questions at issue, but have taken them secondhand from other non-examiners. He concluded that these opinions are 'not worth a brass farthing' (Twain, 1910/2010). The same could be said, however, for many people (including science teachers), and so I intend to treat them a little more seriously than that. With this in mind, it is worth exploring further some of the reasons why people might be anti-science.

Why might people be anti-science?

There is no one answer as to why some people are part of an anti-science/science refuser category. Here I posit just four, but these are intended as illustrative rather than exhaustive.

1. A need for certainty: science is too vague and circumspect

This may seem an odd attitude when we habitues of 'hard' science are so often accused of being overly factual, unerringly exact, painfully pedantic and fixedly inflexible. However, we would certainly agree that science does not have all the answers, has a restricted range of application, is constantly under revision and so is therefore contingent knowledge at best. There are people who need the certainty of correct, concrete answers in life and lose rapid interest if knowledge and understanding is caveated, situated, relational, incomplete. As a science teacher, I have often been confronted by difficult 'why' questions. 'The entire world is made of atoms? Oh really, Mike? Why is that?' I actually do believe that our entire planetary system (and beyond) *is* comprised of atoms and sub-atoms, but arguing from first principles exactly why that is the case can be problematic. An atomic explanation is excellent for describing the here and now of materials, their chemistry and physical properties, but can I be absolutely assured (and therefore assure others) that this is the case for what happens elsewhere? There is physical analysis of 'off-world' meteorites and Moon rocks, of pictures from the robot-rover *Perseverance* on Mars, from the spectral analysis of distant light sources, all providing compelling evidence – but that evidence is heavily dependent on trust in other parts of the story. Moon rocks? But no one has ever been to the Moon, let alone Mars, have they? It was just a film!

There is also some support for the idea that people hold anxieties and fear about uncertainty; they dread the unknown. Which is better as a parent, to put your child at risk of autism or of measles? Which is the greater risk? Can one be absolutely certain of that? A case of measles does not sound so bad; in the old days the theory was to have children catch measles or mumps and get over it as quickly as possible. But polio? Whooping cough? Diphtheria? Meningitis? But what is known about the possible side effects of immunisation? Do they affect all people in the same way? Does 'herd immunity' work? Complete and certain answers allow for a degree of control: the less that is known, the less control is possible. Studies show that unpleasant emotions such as anxiety, as well as uncertainty or feelings of lack of control, contribute to general conspiracy beliefs (Prooijen & Douglas, 2018). Some people need unambiguous answers, control is vitally important, and, for them, science evidence simply does not provide that. As Bertolt Brecht would have it:

> 'I have noticed', said Mr. K., 'that we put many people off our teaching because we have an answer to everything. Could we not, in the interests of propaganda, draw up a list of questions that appear to us completely unsolved?'
>
> (Brecht, 2001)

2.   Science is a conspiracy: 'they' have you in their pocket

There is an old saying: just because you're paranoid doesn't mean they are not out to get you. As a science educator, I really do want people to be critical, sceptical and challenging: do not take everything at face value, do not trust everything you see, hear or read. Do understand that people have vested interests, are subject to commercial pressures, have political ideologies and propaganda agendas. Rather, do follow the evidence, the trends, the patterns, the arguments. Reach a rational, coherent and consistent position for yourself.

Suspicion of, and resistance towards, 'big business', 'big pharma' or 'big government' are understandable, but do science-related commercial conspiracies actually exist? In my view, undoubtedly yes, they do. The tobacco industry denied any connection with lung cancer in the 1950s. They used evidence from asbestos and other scientists to stem the collapse of the tobacco industry by two decades. The makers of Thalidomide were very (very) slow to come to the table. As Goldacre (2012) points out, the 'vitamin industry' and 'food supplement pill industry' are billion-dollar ventures and are certainly prone to exaggeration, misrepresentation, inaccuracy, non-disclosure and distortion. Homeopathy has no basis in science. But I am also with Goldacre when he points out that this is bad science and not science per se. Individuals who tend to see intentional agency behind every event are more likely to believe censorship or conspiracy theories, as are those who attribute extraordinary events to unseen forces or interpret events through the Manichean narrative of good versus evil. There is good evidence, too, that people who have conspiracy theories about one topic will also find home in their thinking for others. So anti-vaccination conspiracies can go hand-in-hand with Big Pharma conspiracies that involve companies manipulating genetically modified crops, medical scientists making spurious connections between HIV and AIDS, scaremongers insisting that human activity exacerbates carbon gas emissions and so on. People who feel this is the way the world works – and for whom this worldview provides a neat way to order, predict and make sense of the universe – might be motivated to believe many conspiracies about science and scientists. Conspiracist ideation is, by definition, difficult to correct because any evidence contrary to the conspiracy is itself considered evidence of its existence (Bale, 2007; Sunstein & Vermeule, 2009). One question, for example, asked of flat-Earthers is: what can possibly be gained from promulgating ideas of a *spherical* Earth? If the Earth is *not* round, then why bother to insist that it is? What possible advantage is to be had by many millions of people, in NASA and elsewhere, conspiring together in order to promote a spherical planet, if in fact it really is flat? One response is that all governments and institutions have something to hide, are seldom honest with the population and that – apparently from common experience – cover-ups are simply the name of the game.

3. Emotions, identity, intuition, superstition: to each their own

Beliefs can be associated with intuitive and non-analytical mental processes rooted possibly in negative emotions and automatic thoughts. When faced with complex judgements, people often quickly and subconsciously ask themselves, 'How do I feel about it?' The resulting feelings serve as a quick and intuitive heuristic that informs their decision – they consider whether something 'feels right' when assessing its accuracy. The sense here is that fears and phobias can underpin some antiscience beliefs. For example, some people have a strong emotional disgust of needles, hospitals, surgery and blood; some gag at the thought of swallowing tablets. Given these heightened reactions, then perhaps more natural remedies appear more palatable – maybe aromatherapy, homeopathy, copper bracelets, healing crystals, burning incense, dreamcatchers, windchimes, horseshoes or rabbit feet. There can be rejections, too, of things thought to be unnatural – herbal remedies are natural, vaccines are toxic.

There are cognitive-motivational styles that favour intuitive, heuristic-driven ways of thinking over more systematic, deliberative modes of processing information, where people put more faith in their ability to use intuition to assess factual claims than in their conscious reasoning skills. Intuition and 'gut feelings' based on everyday experience are important. There are many occasions when decisions must be made on the basis of incomplete knowledge and understanding, where evidence is slim, unavailable or non-existent. On such occasions one has to trust to experience, 'make a judgement call', proceed on 'what feels right'. However, for most people, intuition sits side-by-side with logically considered thinking, sometimes in a tug-of-war. Nor is science devoid of intuition, imagination, analogies, metaphors, 'thought experiments': these are circumstances that can be at the very heart of science: devising imaginative scenarios, generating hypotheses, creating innovative – as yet untested – approaches and suggesting possible alternatives.

4.   Anti-science persuasions

To take a generous view would be to say that anti-science dispositions, such as anti-vaxxers and climate deniers, serve to support a non-conventional identity, the free-spirited non-conformist, the counter-culture individualist, the anti-establishment sceptic, the unique rebel alternative life-styler. This, in turn, might promote an anti-authoritarian perspective where people do not trust accumulating bodies of scientific information or who interpret this information in a way that is opposite to the worldwide consensus, where the educated elite are distrusted – the contrarian hero sometimes called the 'Galileo Gambit'. Great minds in the past have been attacked by the scientific orthodoxy of their day and are invoked by those who want to make individualistic or counter-orthodoxy arguments. The implication is that ideas currently facing close-mindedness and persecution today will eventually be accepted as truth tomorrow. This position is related to the declaration that 'science doesn't have all the answers' or the appeal to 'another way of knowing', where it is alleged that science is not the only source of 'truth'. After all, at its extreme, social constructivism is equated with the assertion that science is only one among many equally valid ways of knowing the world. And that, surely, opens the door to alternative ways of knowing that simply contradict science?

A less generous view might argue that anti-science perspectives rise up from the disagreeable consequences of actually agreeing and accepting, the unpalatable – threatening – implications of having to really do something. So, 'going along' with the science might entail traipsing to a doctor's surgery or a vaccination centre, submitting to a needle, abiding by fussy recycling strictures, paying higher 'green taxes', conforming to mask-wearing regulations or, more dramatically, changing one's unhealthy lifestyle wholesale. Rejection of such actions, and the science behind them, may well also coincide with a lack of concern, of compassion or empathy for others, for the environment, for animals or, for that matter, future generations.

A more antipathetic view, though, sees a conjunction between anti-science, anti-intellectualism and conservative right-wing politics (see, for example, Carrus et al., 2018; Hatzisavvidou, 2021). Debates about science can quickly become political, so that climate science or vaccinations, for example, can be seen as threats to individual freedoms, can curb personal autonomy. Fossil fuels, fracking for oil, gas pipelines (and hence emissions of carbon dioxide) and coal mining all contribute to corporate profit and loss, to standards of living and personal levels of comfort and lifestyle. Rejection of science is strongly associated with a *laissez-faire* view of unregulated free market economics – let people choose; it is the market that dictates what is possible, not an elite cartel of scientists. Gale Sinatra and Barbara Hofer (2021) refer to this as social identity, where people align with others who hold similar beliefs and values. As Naomi Hill's (2021) study shows, political parties and ideologies in the USA are the most influential factors in determining whether someone holds anti-intellectual, anti-science or anti-vax views. That is, people who are left-leaning and centrist in their politics are more likely than their political counterparts to trust the science, be vaccinated, be less concerned with the safety of vaccines and more likely to believe that the benefits of the vaccines outweigh any risks. She adds that ideology and political party are both significant predictors of whether someone also has a fearful disposition towards climate change. There is every reason to believe this is not restricted to one country alone (see, for example, McKinnon (2016) for the UK; Tangney (2019) for Australia; Jylha et al. (2020) for Hong Kong, New Zealand and Sweden).

Back to the debate

So, people are science refusers, are anti-science for a variety of reasons, certainly many more than the four I have outlined here. Adopted in 1948, Article 19 of the Universal Declaration of Human Rights states that: 'Everyone has the right to freedom of opinion and expression; this right includes freedom to hold opinions without interference and to seek, receive and impart information and ideas through any media and regardless of frontiers'. However, as Amnesty International (2021) points out, freedom of speech is *not* the right to say whatever one likes about whatever they like, whenever they like. Freedom of speech and the right to freedom of expression applies to ideas of all kinds, including those that may be deeply offensive. But these freedoms come with responsibilities and can be legitimately restricted. Governments, for example, have an obligation to set out in law ways to prohibit hate speech and incitement. And restrictions can also be justified if they protect specific public interest or the rights and reputations of others. Arguably, anti-vaccination viewpoints and climate change deniers do pose a threat to public health and safety. So, then, should such anti-science/science refusers be abandoned to their own whims and fancies in the name of tolerance and freedom of speech, or should vocal 'anti-vaxxers' and climate deniers, for example, be 'no-platformed' in social settings and media to prevent widespread misinformation? And what is the role of the science educator in all this? Well, yes, science educators commonly work with people young and old to engage and draw them – induct them – into the world of science and, yes, they have varied success in doing so. But, as in so many educational settings, working with the amenable and appreciative is vastly different to working with the awkward, disengaged, testy, truculent and, in some cases, downright mischief-mongers.

In what follows, I set out a series of mini-debates around the roles and directions of science educators.

1.   (In)tolerance

Tolerance is a position of principled self-restraint in the face of behaviours that run counter to the actions and beliefs of the tolerator. Tolerant science educators refrain from interfering with those they dislike or of whom they disapprove, even when they believe that their dislike or disapproval is well-grounded. My own take, though, is that science education should be intolerant. First, by their very nature, anti-science propagators are seldom rational and reasoned in their protestations; their methods are *ad hoc* and unfounded when, for example, 'cherry-picking data'; they seldom apply for research funding or submit their views to professional peer review. Second, their ideas and arguments are commonly strewn with fundamental errors, such as confusing weather events with long-term climate trends. Superstition is defined as a set of beliefs that particular actions can lead to particular outcomes based on mysterious forces rather than scientific knowledge. If and when adverse things do happen, they are written off as chance, bad luck, destiny or fate.

There is a counter viewpoint: to tolerate ignorance as a legitimate choice, a choice not to know, or to know alternatively. As a mundane example, I do not know how many blades of grass are on the rugby pitch of the Principality Stadium in Cardiff. The ground staff at the stadium might (possibly) know, but I have no real need to know, nor am I particularly interested in knowing. I would also defend my right *not* to know. Ignorance conceived in this way is not some passive *lack* of knowledge, far from it; it is active, multidimensional, complex and rarely innocent (Frye, 1983). Understood this way, ignorance is an *activity*, is something that can be performed as a *social practice*, is ritualised. Frye makes the point, too, that ignorance like this is inevitable at both an individual and societal level and is 'entirely in-eradicable' (p. 118). We must simply live with the idea that some people exercise their right not to know, to choose to be ignorant – what the poet John Keats called 'negative capability', the 'capacity for remaining in uncertainties, mysteries, and doubts, without any irritable reaching after fact, logic, and reason' (1899, p. 277).

University teachers tend to work in student-chosen disciplines and so are unlikely to meet many science dissenters; their lecture groups are designed for advocates and habitues. Primary school teachers are also commonly blessed by working with the openly curious and interested. That leaves secondary school science teachers and science educators more widely who will both meet young people (and old) at all four points on my spectrum. While anti-scientists may not have the unalienable right to proselytise the views they hold at all times in all contexts, they would certainly (in my view) have the right to construct their own ignorance, and to remain squarely within it. The best, I guess, we science educators can do is to remain stoically resilient and continue, Sisyphus-like, to chip away at ignorance wherever we find it.

2.   The implicit and the explicit

When science habitués like me encounter science refusers like these, our common response is to repeat the evidence as clearly and deftly as possible, defend facts and principles and use rational, logical debate to defeat misinformation, misconceptions and downright misrepresentation. Personally satisfying, maybe, but it is commonly a wholesale waste of time and energy. For example, repeating evidence and counterevidence has been shown to have little to no effect on hard-boiled convictions and beliefs. There are studies to suggest, too, that fighting 'myths with facts' might actually backfire – unintentionally cementing the very ideas one intends to challenge. A study of Californian parents found that countering anti-vaccination misconceptions with clear explanation was actually possible, but that the use of such a rational approach with parents resolutely opposed to vaccination simply further entrenched their ill-founded views (Nyhan et al., 2014). While I might feel that evidence drives attitudes, this is clearly not always the case. In many circumstances, people develop attitudes through emotions, intuitions and 'gut feelings' in ways that are difficult to articulate, and then search for evidence to explicate, support, reinforce and confirm these. Rational argument is often powerless in the face of stubborn ideology, faith, trust and belief.

So, I commonly begin conversations agreeing with people that science does not necessarily have all the answers or even good answers – there are many things unknown to humankind, new things are discovered every day, theories constantly change and so on. It is sometimes a difficult line to be walked: I am quite clearly taken as something of an expert while, at the same time, downplaying that expertise. Nor is that just the case with adults; I have always found that asking children credulous questions beats telling – and adults' and children's own questions trump my questions any day (Pedrosa de Jesus & Watts, 2019). In doing so, there is a need to make the implicit explicit and to recognise the role of emotions. One aim of science education, and therefore an important role for all science educators, is to enable scientifically literate adults to understand, for example, media reports, make informed personal decisions (e.g., about diet or medical treatment) and more generally engage with the issues posed by science and technology. One aspect of this aim relates to Gert Biesta's notion of 'subjectification', which I translate here as 'being equipped to join the conversations of science'. This equipping is less to do with science subject knowledge *per se* as it is more about ways of working which open opportunities for people, young and old, to make contributions and to 'find their science voices'. Science educators should, in my view, develop both understanding and skills in their charges, resulting in Michel de Montaigne's *'mieux vaut une tête bien faite que bien pleine'* ('A brain well-formed rather than one well filled', cited in Hall, 1997, p. 61).

Now, it will certainly seem that 'enabling people to find their science voice' is a long way from the 'eradication of ignorance' with which I began, but I begin with a positive thrust. My own research work over many years has explored public understanding of science in several different arenas (Salehjee & Watts, 2021; Watts, 2015), and through numerous science discussions with everyday people from all walks of life. I have called these discussions 'kitchen conversations', not least because the majority have actually taken place over coffee in my respondents' kitchens or sitting rooms. They very often begin by people shying away from science, as in, 'Oh, I was never very good at science in school, it just wasn't for me'. Throughout our conversations, though, it is quite clear that they do have a clear interest once they begin to see that science is broad, wide and nowhere near as esoteric as they initially imagined. They have views about diet and exercise, air pollution, garden pests, plastic pollution, extracting fuel from oil wells, personal health issues, volcanic eruptions, Ash Dieback disease (*Hymenoscyphus fraxineus*), driverless cars, the causes of cancer, the decline in bee populations, space travel and much more besides. They begin by expressing their views tentatively and often punctuated by 'Oh, I was watching this TV programme . . .', 'I read somewhere that . . .' or 'My neighbour was telling me . . .', but, once in the flow, they find that they do indeed have a 'science voice'. And, once into these conversations, I do the same. Describing myself as a 'science habitué' does not make me a polymath, and I am happy to admit that, while I have some knowledge in depth, I have only a skittish, superficial, potpourri, 'quiz night' grasp of many other areas of science, so I, too, often resort to 'popular science'.

And, like Gale Sinatra (2021), I appeal to my conversationalists' many social identities: being parents, grandparents, carers, to make a connection and the case for immunisation or recycling. Moreover, I am mindful of those science refuse-niks out there and counsel myself against indulging in intellectual arrogance: 'I know and you don't know, and I'm here to tell you how it is'. To confront an uninformed, unpopular or offensive opinion as being arrant nonsense 'is to assume our own infallibility', as John Stuart Mill once said in *In Defence of Liberty*

(1859). I have no wish to add to what Herbert Marcuse described as 'the systematic moronisation of children and adults alike'.

3.   Start early, stay late

In their 2013 paper, Louise Archer and colleagues say, 'Students . . . who do not express STEM-related aspirations at age 10 are unlikely to develop STEM aspirations by the age of 14 when they are making important subject choices for their GSCEs'. If that is indeed the case, then it places a heavy onus on primary school education, not necessarily to turn all pupils into scientists but, at least, to hold the open the door to science and help not to slam it shut. That said, there is also ample evidence to show that young people turn off science very rapidly at the onset of secondary schooling. Some two decades ago, Robin Millar and Jonathan Osborne (1998) wrote that:

> School science, particularly at secondary level, fails to sustain and develop the sense of wonder and curiosity of many young people about the natural world. This interest and inquisitiveness which characterises many primary school children's responses to science diminishes at secondary school to a degree that cannot be wholly be accounted for by the onset of adolescence.
>
> (p. 5)

Not much (in my view) seems to have changed in the intervening years (Salehjee & Watts, 2021). Nor have efforts at improving general, long lamented, 'science literacy', or the public understanding of science, fared much better (Hazen & Tefil, 2009). Christoph Henseler and Hans-Liudger Dienel (2017) maintain that the general public's lack of basic science is sometimes attributed to the poor communication skills of scientists, not least because, as a former head of the National Science Foundation claimed:

> With the exception of a few people . . . we don't know how to communicate with the public. We don't understand our audience well enough. . . . We don't know the language and we haven't practiced it enough.
>
> (Weigold, 2001, p. 38)

Justin and his colleague Lucy Avraamidou (Dillon & Avraamidou, 2021a, 2021b) are much more emphatic: they make the strong point that current science education has been an outright failure. It has failed not just, as Archer says here, in developing curiosity and wonder, but in its wholesale (under)development of science literacy in the population. In a nutshell, successful science education would really not leave the vast majority of the public inadequately prepared for a pandemic or for the threat of climate change. Successful science education would mean that politicians of all colours would be able to follow the science of Covid-19 with a good understanding of basic scientific concepts and trust in how science works. Dillon and Avraamidou call for radical systematic reform of schooling:

> Young people do not just need more or different knowledge, they need a fundamentally different educational experience. The nature and purpose of education needs to be re-envisioned to address contemporary societal challenges, and it needs to be done urgently. . . . Unless we rethink what it means to be scientifically and environmentally literate within the broader social, political and economic context, then science education will continue to fail. If students emerge from 11 or more years of schooling without understanding how science works, then any claim that we are providing an effective science education is nonsense.
>
> (2021a)

So, the debate is clear: a necessary programme for the eradication of science ignorance entails a radical rethink and wholesale redesign of science education (in schools, universities and elsewhere). Have no truck at all with the idea that ignorance is a valid choice. Thank you, Justin, thank you, Lucy; there's no more fitting way to highlight where debate lies at the very heart of science education. We are certainly *not* at one with the editor of *New Scientist*, whom Richard Dawkins famously quoted (YouTube, 2013) as saying, 'Science is interesting and if you don't agree you can f**k off.'

Questions for further debate

1   How far should we go in revisioning science education?
2   Should teachers holding anti-science views be allowed to teach in schools?

Suggested further reading

Al-Khalili, J. (2017). *What's next? Even scientists can't predict the future – or can they?* Profile Books.
Brockham, J. (2015). *This idea must die. Scientific theories that are blocking progress.* Harper Perennial.

References

Amnesty International. (2021). Free Speech. https://www.amnesty.org.uk/issues/free-speech
Archer, L., Osborne, J., DeWitt, J., Dillon, J., Wong, B., & Willis, B. (2013). *ASPIRES young people's science and career aspirations, age 10–14.* King's College London. https://www.stem.org.uk/resources/elibrary/resource/116495/aspires-young-peoples-science-and-careers-aspirations-age-10-14#&gid=undefined&pid=1
Bale, J. M. (2007). Political paranoia v. political realism: On distinguishing between bogus conspiracy theories and genuine conspiratorial politics. *Patterns of Prejudice, 41*, 45–6.

Brecht, B. (2001). *Stories of Mr. Keuner* (trans. M. Chalmers). City Lights.

Carrus, G., Panno, A., & Leone, L. (2018). The moderating role of interest in politics on the relations between conservative political orientation and denial of climate change. *Society and Natural Resources, 31*, 1103–1117.

Dawkins, R. (2013). *If you are not interested in science*. https://www.youtube.com/watch?v=-Fh_liyhIH8

Dewey, J. (1922). *Human nature and conduct: An introduction to social psychology*. Henry Holt.

Dillon, J., & Avraamidou, L. (2021a). Science education has failed. *Education in Chemistry*. https://edu.rsc.org/opinion/science education-has-failed/4013474. article

Dillon, J., & Avraamidou, L. (2021b). Towards a viable response to COVID-19: Questions for the science education community. *Education in Science, 283*, 20–21.

Frye, M. (1983). *The politics of reality: Essays in feminist theory*. Crossing Press.

Goldacre, B. (2012). *Bad pharma: How medicine is broken and how we can fix it*. HarperCollins Publisher.

Hall, M. L. (1997). Montaigne's uses of classical learning. *Journal of Education, 179*(1), 61.

Hatzisavvidou, S. (2021). 'The climate has always been changing': Sarah Palin, climate change denialism, and American conservatism. *Celebrity Studies, 12*(3), 371–388.

Hazen, R. M., & Tefil, J. (2009). *Science matters. Achieving scientific literacy*. Anchor Books.

Henseler, C., & Dienel, H.-L. (2017). Maps of the uncertain: A new approach to communicate scientific ignorance. *Innovation: The European Journal of Social Science Research, 30*(1), 121–136.

Hill, N. (2021). *Anti-intellectualism and American fears: An analysis of social and political factors that influence distrust in scientific authority*. Chapman University COVID-19 Archives, Chapman University Commons.

Jylha, K. M., Tam, K.-P., & Milfont, T. L. (2020). Acceptance of group-based dominance and climate change denial: A cross-cultural study in Hong Kong, New Zealand, and Sweden. *Asian Journal of Social Psychology, 24*(2), 198–207.

Keats, J. (1899). *The complete poetical works and letters of John Keats, Cambridge Edition*. Houghton, Mifflin and Company.

McKinnon, C. (2016). Should we tolerate climate change denial? *Midwest Studies in Philosophy, 40*(1), 205–216.

Mill, J. (1859). *In defence of liberty*. East India Publishing Company.

Millar, R., & Osborne, J. (1998). *Beyond 2000. Science education for the future*. Nuffield Foundation.

Nyhan, B., Reifler, J., Richey, S., & Freed, G. L. (2014). Effective messages in vaccine promotion: A randomized trial. *Paediatrics*. http://pediatrics.aappublications.org/content/early/2014/02/25/peds.2013-2365.abstract

Pedrosa de Jesus, M. H., & Watts, D. M. (2019). *Academic growth in higher education: Questions and answers*. BRILL/Sense Publications.

Prooijen, J. W., & Douglas, K. (2018). Belief in conspiracy theories: Basic principles of an emerging research domain. *European Journal of Social Psychology, 48*, 897–908. https://doi.org/10.1002/ejsp.2530

Salehjee, S., & Watts, D. M. (2021). *Becoming scientific. Developing science across the life-course*. Cambridge Scholars Publishing.

Sinatra, G. M. (2021). Motivational and emotional impacts on public (mis)understanding of science. *Educational Psychologist*, 1–10.

Sinatra, G. M., & Hofer, B. J. (2021). *Science denial: Why it happens and what to do about it*. Oxford University Press.

Sunstein, C. R., & Vermeule, A. (2009). Conspiracy theories: causes and cures. *Journal of Political Philosophy*, *17*, 202–227.

Tangney, P. (2019). Between conflation and denial – The politics of climate expertise in Australia. *Australian Journal of Political Science*, *54*(1), 131–149.

Twain, M. (2010). *Autobiography of Mark Twain, the authentic original version*. Seven Treasures. (Original work published 1910)

Watts, D. M. (2015). Public understanding of plant biology: Voices from the bottom of the garden. *International Journal of Science Education*, *5*(4), 339–356.

Weigold, M. F. (2001). Communicating science: A review of the literature. *Science Communication*, 23(2), 164–193. https://doi.org/10.1177/1075547001023002005

Debates about the relationship between science and science pedagogy

Chapter 8

Challenges in teaching using inquiry-based science

Robyn M. Gillies

Introduction

Emphasis in recent years has been on teaching science using an inquiry approach, where students are active participants in investigations that challenge their curiosity, encourage them to ask questions, explore potential solutions, use evidence to help explain different phenomena, debate alternative positions, and predict potential outcomes if different variables are controlled (Duschl & Grandy, 2008). The inquiry process is complex, as it involves the students reconciling their current understandings of a problem with both the evidence obtained from an inquiry and the ability to communicate their newly acquired knowledge in a way that will be accepted as well as reasoned, logical, and justifiable. In essence, Llewellyn (2014) maintains that inquiry refers to a variety of processes and ways of thinking that contribute to the development of new understandings and knowledge in science. It not only involves students in 'doing science', but it also includes the processes scientists use to develop new understandings and knowledge about the nature of science.

Scientific inquiry recognizes the diverse ways in which scientists study the natural world, evaluate evidence, and propose explanations about the phenomena under investigation. It also refers to "the activities through which students develop knowledge and understandings of scientific ideas, as well as understandings of how scientists study the natural world" (National Science Teachers Association, 2004, p. 1). To understand science, students need to have opportunities to do science by participating in activities, completing projects, investigating topics that challenge their curiosity, and discussing their research and readings. As students learn to emulate how scientists undertake their inquiries, they obtain a better understanding of the processes involved in conducting scientific investigations. In so doing, they learn how to pose and answer questions on topics that they are investigating, and to differentiate between evidence and inference (Bybee & Van Scotter, 2007).

The 5E instructional model

One approach to teaching inquiry science that has a strong evidence base is the 5E's instructional model proposed by Bybee (2014, 2015). This model

DOI: 10.4324/9781003137894-10

consists of five phases of learning: *engage, explore, explain, elaborate*, and *evaluate*. In the first phase, teachers begin by capturing students' attention and interest by engaging them in tasks that challenge their curiosity. This may be a short video, role play, event, or problem scenario requiring students to confront the cognitive disequilibrium they experience and to seek clarifications on the dilemma they are confronting. Bybee maintains that asking questions, posing problems, or presenting conflicting scenarios are examples of strategies that *Engage* students' attention and interest. These strategies also present opportunities for teachers to informally identify misconceptions expressed by the students and provide opportunities to investigate the topic in more depth. Once the students have had a chance to *Explore* the phenomena and clarify their misconceptions, they are encouraged to *Explain* how they resolved the dilemma so others in their class have opportunities to reconcile the explanation with their own understandings or to rebut it. By using the students' explanations and experiences as the context, the teacher is then able to introduce the scientific or technological concepts associated with the problem, enabling the students to gain a clearer understanding of key issues, vocabulary, and practices. The teacher's role during this phase is to provide opportunities for students to become involved in learning experiences that extend and expand on the previous phases, so they develop more depth and scope in their understandings of the phenomena that they are investigating. During this phase, the teacher encourages students to use each other as a resource, as well as other sources of information such as web-based searches, written texts, and simulations as they seek to *Elaborate* on the scientific concepts under investigation.

In this phase, the students participate in learning experiences that expand and enrich the concepts and understandings that they developed in previous phases to new situations that are closely aligned to what they know and understand. The emphasis in this phase is on students being able to generalize information and understandings to new situations or experiences. Interaction is critically important as students share their insights and conceptions with others through writing reports, producing portfolios, participating in debates that challenge current conventions, or utilizing different graphic modes to present information that provide additional insights on the topic at hand.

Evaluate is the final phase, and involves students receiving feedback on the suitability of the explanations they have provided. This can be done quite informally in a recursive and cyclic manner as they work on different phases of their inquiry. For example, questions such as: *What did you think about the explanations provided? On the one hand, I hear you saying that . . . but on the other you're telling me. . . . How do you reconcile these perspectives?* Students can also receive feedback in a more formal manner through criterion referenced assessment tasks, such as group projects, portfolios, exhibitions of performances, and work samples, which may include essays, video-presentations, or creative art works (Gillies, 2009).

Do inquiry-based science approaches work?

The evidence for inquiry-based science being able to motivate students' interest in science, and, in turn, enhance their approach to learning science is fairly consistent. Chinn et al. (2013) argued that inquiry methods, while time-consuming, have the advantage of promoting both content understanding and growth in reasoning. This occurs because during inquiry-based instruction, students are challenged to develop explanatory conceptions of their understandings in a particular domain (e.g., theories on different states of matter) and engage in the processes of reasoning from evidence. In fact, Chinn et al. argue that in inquiry-based instruction, conceptual change often involves changes in both explanatory conceptions and standards and practices of reasoning.

In a meta-analysis of 15 published journal articles on inquiry-based learning in science, Firman et al. (2019) found that inquiry-based learning had a mean effect size of 0.45, indicating that this approach to teaching science had a positive effect on improving students' mastery of science concepts, the acquisition of epistemological beliefs, and motivation to learn. The results led Firman et al. to conclude that inquiry-based learning can be used to promote students' development of inquiry skills in primary and high school settings.

Treagust et al. (2020) reported on a student-centred intervention called Process-Oriented Guided Inquiry Learning (POGIL) in a sample of 122 students in Grade 10 from two male government schools in Qatar. The study sought to determine students' perceptions of their learning before and after the POGIL intervention and whether this approach to teaching science was meeting the expected learning outcomes of the students' chemistry curriculum. The results showed that when surveyed on their perceptions of what is happening in this class (pre- and post-intervention), the intervention group differed significantly from the control group on their positive perceptions of student cohesiveness, involvement, cooperation, personal relevance, enjoyment of chemistry lessons, and academic efficacy. The intervention group also recorded significantly higher achievements in learning chemistry at post-intervention than their peers in the control group.

Kang and Keinonen (2018) examined a large-scale data set from the Program for International Students Assessment (PISA) 2006 and reported on the effect of four student-centred approaches to learning on students' interest and achievements in science. The authors found that topic-based approaches and guided inquiry-based learning were strong positive predictors for students' achievement in science, and they were also positively associated with students' interest. In contrast, open inquiry-based learning and discussion-based learning were strong negative predictors of students' achievement and interest in science. The authors concluded that students become more interested in science when they have opportunities to participate in more guided-inquiry learning, particularly when connections are made between their school science experiences and real-life experiences. Furthermore, this increased interest in science accounts for better achievement outcomes.

Similar results were reported by Areepattamannil et al. (2020) in a large international study of 428,197 adolescents from 15,644 schools in 66 countries that analyzed data on adolescents' dispositions towards teacher-directed versus inquiry-based science instruction. Interestingly, the results showed that both teacher-directed and inquiry-based science instruction are significantly positively related to adolescents' enjoyment of science, interest in broad science topics, motivation to learn science, and science self-efficacy. The authors suggested that a blend of teacher-directed and inquiry-based science instruction may be more appropriate for developing students' positive dispositions towards science.

In summary, recent research (Treagust et al., 2020; Areepattamannil et al., 2020) shows that students do learn when they participate in inquiry-based science experiences. Changes have been reported in explanatory conceptions, mastery of science concepts, and changes in standards and practices of reasoning. Additionally, students are often more motivated to learn science, possibly because of the hands-on approaches to learning science that are adopted, leading students to develop more positive attitudes towards science and better achievements in science (Kang & Keinonen, 2018).

Challenges in implementing inquiry

Given that the evidence has been well documented that achievement and affective benefits accrue to students when they have opportunities to participate in inquiry-based learning in science, it seems incongruous that teachers continue to teach science using transmission approaches. One study that has investigated teachers' perceptions of teaching inquiry-based science was conducted by Gillies and Nichols (2015), which drew on the experiences of nine Grade 6 teachers from five different schools who taught two units of inquiry-based science across two school terms (see Gillies et al., 2012). In particular, the study focused on the teaching processes they employed and the benefits and challenges of implementing cooperative learning, a student-centred approach where students worked in small teams to investigate topics. All teachers spoke positively about teaching the inquiry-based science units because the inquiry topics not only captured students' interest, but the process also allowed them to take ownership over how they learned. The teachers perceived that cooperative learning acted as a driver for the inquiry, allowing students to exercise responsibility for what the group learned and to think outside the box when needed. However, the teachers did emphasize the importance of structuring the inquiry process to challenge students' thinking and scaffold their learning to encourage discussion.

Similar results were reported by Tseng et al. (2013), who interviewed experienced junior high school teachers to gauge their perceptions of teaching science by inquiry. Just as the teachers in the Gillies and Nichols (2015) study believed it was important to structure and scaffold the students' learning, Tseng et al. also found that the teachers in their study reported that it was important to structure and scaffold the inquiry experiences if students were to acquire the knowledge and skills needed to engage successfully in learning. This included using different strategies that were well-structured, where students worked through a process to arrive at a solution to ill-structured strategies where the topics were more open and discovery-based. The former is designed to ensure that students master specific steps in solving problems while the latter requires students to interact together as they search for solutions to problems at hand. Being adaptable and able to use different inquiry strategies interchangeably, depending on students' needs, were seen as important if teachers were to become proficient at teaching inquiry.

There is no doubt that teachers need support if they are to be flexible and adaptable in the strategies that they use to teach inquiry-based science (Areepattamannil et al., 2020). Given the propensity of many teachers to continue to use transmission approaches to teach where the discourse is uni-directional rather than multi-directional, understanding how to utilize different strategies for teaching inquiry remains a challenge (Howe & Abedin, 2013). The following discusses three associated curricula strategies that are critically important in implementing inquiry-based science: cooperative learning, scientific literacy, and scientific discourse.

Cooperative learning

One approach to learning that underpins inquiry is cooperative learning, where students work in small groups on problem-based tasks. While meta-analyses consistently document the academic and social benefits that students derive from working cooperatively together (Johnson et al., 2014), successful group work is very dependent on how the groups are structured and the types of tasks they undertake. It is well recognized that placing students in small groups and expecting them to know how to work together will not necessarily promote cooperation. In fact, groups often struggle with knowing what to do and, in the process, discord can occur as members grapple with the demands of the task, as well as managing the process involved in learning, including how to deal with divergent opinions or working with peers who make minimal contributions to the group (Gillies, 2007).

Research has identified five key elements that need to be embedded in groups in order for members to cooperate (Johnson & Johnson, 2002). These elements are:

1   Positive interdependence where members perceive that they are linked together in such a way that in order to achieve their goals, they must assist others to achieve theirs as well. Positive interdependence can be structured in groups so that each member has to complete part of the task, for example, a group project where students work on one part of the larger task. Members then share their individual tasks or small group tasks (e.g., two members

working together) with others in the group to complete the larger group task. Interestingly, Johnson and Johnson found that when group members have small tasks that they must complete in order for the group to be able to complete its larger task or goal, group members will map on to this experience the perception of psychological interdependence, and it is this state that provides the momentum for members to cooperate. It is essentially a case of sink or swim together.

2   Promotive interaction occurs when group members realise that they need to discuss the task with each other, share ideas and information, and work constructively to find a solution to the problem at hand. Teachers have a responsibility to ensure that group members understand that it is important that they acknowledge the efforts of others, facilitate access to information and resources, and be 'tuned in' to the needs of different group members. When group members demonstrate these types of behaviours, students are more likely to feel accepted and valued, less anxious and stressed, and more caring and supportive of each other.

3   Social skills are evident when group members are respectful of each other, actively listen to what others say, consider the different perspectives that members bring to the group, constructively critique ideas and clarify differences or misconceptions, and accept responsibility for one's own actions or behaviours. Other very important skills include taking turns in expressing opinions or sharing resources or roles, sharing tasks fairly so no member is overloaded with work, and engaging in democratic decision-making processes.

4   Individual accountability involves group members realizing that it is important for all members to complete their part of the task while encouraging others to do likewise. The acceptance of individual accountability helps to build group cohesion and motivation as group members realize the importance of their contributions to achieving the group's goals. Individual accountability can be established in groups by assigning different tasks or roles to members that need to be completed if the group is to achieve its goal.

5   Group process and reflecting enables group members to discuss how well they worked to achieve the group's goal and what they may still need to do. These are processes that are critically important for student learning, as Johnson and Johnson (2009) found that they promote greater success in problem-solving and achievement gains than students in cooperative groups who do not follow up with processing their experiences in the group.

Scientific literacy

If students are to understand how science can be used as a way of thinking, finding, organizing, and using information to make decisions, it is critically important that they are scientifically literate. Scientifically literate people are interested in the world in which they live, engage in discussions about scientific issues, interrogate the evidence on claims made by others, and draw on evidence to make informed decisions about different phenomena that affect their world (Rennie, 2005). In a similar vein, Krajcik and Sutherland (2010) note that for students to become scientifically literate, they need to be able to develop an understanding of science content and scientific practices, critique the evidence, and participate in making decisions that affect them personally or others in the community.

Krajcik and Sutherland (2010) have identified five instructional and curricula features that can support students in developing scientific literacy. These features include:

> (i) linking new ideas to prior knowledge and experiences, (ii) anchoring learning in questions that are meaningful in the lives of students, (iii) connecting multiple representations, (iv) providing opportunities for students to use science ideas, and (v) supporting students' engagement with the discourses of science.
>
> (p. 456–457)

The following illustrates how these five literacy features can be implemented:

1   Linking new ideas to prior knowledge and experiences involves teachers contextualizing the content to be taught with students' previous knowledge and experiences. Giamellaro (2014) found that when teachers did this, students' conceptual knowledge of science was boosted, leading to more positive learning outcomes. Furthermore, students' interest in science was enhanced, as they were better able to understand its relevance to their daily lives. However, Giamellaro also noted that teachers need to be mindful of the importance of carefully facilitating students' learning to help them understand the relative value of the new information they are acquiring.

2   Asking questions that are meaningful and important in the lives of students will help to challenge their curiosity and motivate them to investigate the topic under discussion. King (1997) identified a series of question stems that could be used to elicit different types of thinking in students. For example, (a) review questions tend to ask students to recall information; (b) probing questions ask students to consider the possibilities that could occur; (c) hint questions challenge students to consider the consequences; (d) thought-provoking questions are designed to challenge students to consider different perspectives; and (e) metacognitive questions are designed to encourage students to think about the thinking they have undertaken to solve a problem. In science lessons, these types of questions can be used interchangeably to gauge students' understandings while challenging and scaffolding their thinking, enabling the teacher to assess how the students are managing the content and processing the information they receive (Gillies et al., 2014).

3   Connecting multiple representations involves not only teaching students how to interpret different visual symbols in text, graphs and tables,

simulations, and graphical representations, but also to understand different aural and embodied representations that are used to support meaning. In fact, Tang and Moje (2010) argue that students need to be exposed to both multiple representations and multimodal representations concurrently. In the former, students are introduced to the same concept through different representational forms, whereas in the latter, they learn to build conceptual understandings through the simultaneous use of different modalities (e.g., visual, aural, tactile) within and across different representations. Tang and Moje see these two processes as interactive and complementary as students learn to interpret and construct meaning from exposure to multiple modes of representation.

4   Providing opportunities for students to use science ideas includes helping students to discuss, represent, critique, and apply their developing understandings about the situations or new problems they are confronting. Teachers play a critical role in checking that they scaffold and challenge students' emerging understandings to ensure that their interpretations and reasoning about the phenomena under investigation are aligned with their experiences, data, or available supporting evidence (Gillies, 2016). Murphy et al. (2019) reported on the effect that teachers' participation in a targeted inquiry-based/nature of science programme continuing professional development program had on primary students' experiences of scientific inquiry and developing conceptions of the nature of science. The results showed that not only did teachers participation have positive effects on children's experiences of scientific inquiry and their developing conceptions of the nature of science, but it also led to a significant increase in students' engagement with student-led inquiry-based approaches and the development of more elaborate conceptions of the nature of science.

5   Supporting students' engagement with the discourses of science, where students learn how to communicate their understandings, listen to others and build on their ideas, seek additional clarification and information when needed, construct explanations, and learn how to critique propositions that may need to be challenged. Teachers model many of these ways of talking about science when they introduce students to a topic, seek students' participation in discussing it, and engage in dialogic exchanges where teachers and students build on each other's ideas by adding additional comments to construct coherent and logical lines of inquiry. This type of discourse, Krajcik and Sutherland (2010) argue, is essential if students are to learn how to "talk and write about science and to practice supporting their ideas with evidence" (p. 458). Teaching students to develop arguments that are based on evidence is certainly a challenge that teachers confront as they must decide on whether they are going to teach students how to engage in persuasive arguments or a consensus building approach that uses reasoned discussion (Resnick et al., 2010).

Felton et al. (2009) reported on a study in which 100 seventh-grade students were randomly assigned to one of three conditions: a disruptive (persuasive) condition; a deliberative (consensus) condition; or a control condition. Their task was to discuss specific socio-scientific issues on climate change across eight 50-minute science lessons. The results showed that argumentative dialogues can improve content learning and argument quality, with the students in the deliberative condition obtaining significantly higher results than students in the control condition, while the students in the disputational condition fell in between. The authors argued that argument of any sort (deliberative or disputational) had a positive effect on students' reasoning skills, although students in the deliberative condition constructed arguments that demonstrated greater attention to claims and evidence on both sides of the issue.

A follow-up study by Felton et al. (2015) also found that persuasion versus consensus while arguing can affect both students' acquisition of content and reasoning. Again, the results showed that when students had opportunities to engage in using argumentative discourse in science where they were expected to reach consensus on a topic, they were more likely to use moves that elicited, elaborated on, and integrated their partners' ideas than students who tried to persuade or defend their arguments with their peers. Felton et al. argued that teachers need to be mindful of the goals and purposes of the discourse; persuasion is adversarial with speakers advancing incompatible claims, while a consensus approach involves seeking to resolve incompatible claims through collaborative discussion, and decision-making is more likely to lead to the co-construction of knowledge and understandings.

In short, if students are to become scientifically literate, teachers need to ensure that they have opportunities to participate in discussions with each other about scientific issues where they learn to affirm or challenge ideas, present evidence to support any claims that they make, and engage in the process of making evidence-based arguments. Constructing explanations and arguments are essential components of scientific discourse, and it is critically important that students have opportunities to use these ways of talking to illustrate how they reasoned from the evidence that was available to them (Krajcik & Sutherland, 2010). Moreover, teachers are more likely to engage students in reasoned discourse when they model the importance of disciplinary knowledge and well-structured reasoned discussions (Resnick et al., 2010).

Scientific discourse

Emphasis over the last three decades has been on encouraging teachers to move from teacher-centred approaches to teaching to encouraging students to actively participate in class discussions where they can share ideas, information, and understandings with their teacher and peers. This has happened because there is an enormous volume of research that attests to the benefits, both social and academic, that students derive when they have opportunities to interact with others on problem-based topics that challenge their thinking and understandings. This type of teaching is referred to as dialogic teaching, where teachers utilize the power of talk to engage students' interest and extend their thinking while building on their ideas to develop new knowledge and understandings (Alexander, 2018). Teachers who engage in dialogic teaching recognize that teaching involves the following key criteria:

(a) It is collective. Teachers and students collaborate to address problem issues. This can be achieved in small groups or in the larger class.

(b) It is reciprocal. Students and teachers listen to others' ideas, share information and ideas, and consider different perspectives on a topic or problem.

(c) It is supportive. Students feel free to share their ideas without being sanctioned or criticized when others disagree with their ideas. In such an environment, students and teachers work constructively together to resolve problems issues.

(d) It is cumulative with teachers and students building on their own and others' ideas to develop logical lines of thinking and inquiry.

(e) It is purposeful with teachers guiding the students' talk to pursue specific goals.

When the teacher actively engages in dialogic teaching, students acquire a repertoire of oracy, based on the modelling and scaffolding that the teacher has elucidated that helps them to interact with each other. Alexander (2018) refers to this as the "learning talk repertoire" (p. 104) and it includes the abilities to narrate, explain, ask questions that probe and challenge, elaborate on responses, argue, reason and justify, speculate and imagine, and negotiate. This talk repertoire, though, is contingent on students being prepared to listen to each other, consider different perspectives, reflect on what they hear and learn, and give others time to think before proposing an idea or solution. In a sense, there is a careful balance between students and students and students and teachers' dispositions to be respectful of each other's ideas, perspectives, and ways of talking as they work together to find solutions to different problems.

Alexander (2018) reported that when dialogic teaching has been embedded in classroom discourse and students have learned to utilize the learning talk repertoire, the following changes were evident in classroom interactions:

1   Teachers were asking more probing questions designed to encourage students to think and speculate about the topic under discussion.
2   Student-teacher exchanges were longer and more substantive as each built on the ideas or information presented.
3   Teachers were more student-centered as they prompted and facilitated students' responses.

4   Students were speaking more readily, clearly, and audibly.
5   More students were initiating interactions by asking questions, making suggestions, and commenting on others' responses.

In the science classroom, the teacher's role is to establish the conditions that not only capture students' interest in the topics to be discussed, but also ways of interacting that promote and facilitate dialogic exchanges. Teaching science, Huff and Bybee (2013) maintain, involves providing students with opportunities to be involved in the dynamic exchange of ideas, build connections between ideas and evidence, and use evidence to support or rebut propositions. When teachers create opportunities for students to engage in these types of dynamic exchanges, they develop the skills of argumentation as they learn to make claims and justify their scientific understandings while identifying both the strengths and shortcomings in others' arguments. With practice, students develop critical discourse where they learn to engage in argumentation that helps them to strengthen their conceptual understandings and scientific reasoning capabilities.

In a study that investigated the role of argumentative discourse in middle-year inquiry science classes, Bathgate et al. (2015) found that it had a positive effect on students' capacities to think critically, to reflect on their ideas and elaborate on them, and to evaluate their initial misconceptions in the light of new evidence. In a similar vein, Soysal (2021) reported on a study that investigated the relationship between the implementation of an argument-based inquiry approach in science, teachers' talk moves, and students' critical thinking in middle-years science classes. The results showed that argument-based inquiry and teachers' talk moves, particularly those that challenged students' thinking, invited them to present their data-based interpretations, and encouraged them to communicate with each other, stimulated their higher-order critical thinking. In sum, Bathgate et al. and Soysal found that when students are taught how to engage in argumentative discourse during inquiry-based science, it had a positive effect on their higher-order critical thinking capacities.

Conclusion

This chapter has highlighted the importance of structuring inquiry-based science activities so that students are actively involved in investigations that challenge their curiosity, encourage them to ask questions, explore possible solutions, use evidence to help explain different phenomena, and understand how to engage in argumentative discourse. There is no doubt that the inquiry process is complex, as it involves students in not only considering the evidence and reconciling it with their current understandings, but also being able to communicate their newly acquired knowledge to others in ways that are accepted as logical and well-reasoned. The teacher's role is pivotal to not only constructing inquiry-based science experiences that will help students develop an understanding of the scientific content, but also the practices that enable them to engage in well-reasoned discussions that facilitate critical thinking and learning.

Questions for further debate

1   How and when should investigations be used in science teaching?
2   How can the effectiveness of inquiry-based approaches be evaluated?

Suggested further reading

Lee, E. A., & Brown, M. J. (2018). Connecting inquiry and values in science education. *Science & Education, 27*, 63–79.
Van Uum, M. S. J., Verhoeff, R. P., & Peeters, M. (2017). Inquiry-based science education: Scaffolding pupils' self-directed learning in open inquiry. *International Journal of Science Education, 39*(19), 2461–2481.

References

Alexander, R. (2018). Developing dialogic teaching: Genesis, process, trial. *Research Papers in Education, 33*(5), 561–598.
Areepattamannil, S., Cairns, D., & Dickson, M. (2020). Teacher-directed versus inquiry-based science instruction: Investigating links to adolescent students' science dispositions across 66 countries. *Journal of Science Teacher Education, 31*, 675–704.
Bathgate, M., Crowell, A., Schunn, C., Cannady, M., & Dorph, R. (2015). The learning benefits of being willing and able to engage in scientific argumentation. *International Journal of Science Education, 37*, 1590–1612.
Bybee, R. (2014). The BSCS 5 E instructional model: Personal reflections and contemporary implications. *Science and Children, 51*(8), 10–13.
Bybee, R. (2015) *The BSCS 5 E instructional model: Creating teachable moments.* National Science Teachers' Association Press.
Bybee, R., & Van Scotter, P. (2007). Reinventing the science curriculum. *Educational Leadership, 64*(4), 43–47.
Chinn, C., Duncan, R., Dianovsky, M., & Rhinehart, R. (2013). Promoting conceptual change through inquiry. In S. Vosniadou (Ed.), *International handbook of research in conceptual change* (2nd ed., pp. 539–559). Routledge.
Duschl, R., & Grandy, R. (2008). Reconsidering the character and role of inquiry in school science: Framing the debates. In R. Duschl & R. Grandy (Eds.), *Teaching scientific inquiry: Recommendations for research and implementation* (pp. 1–37). Sense.
Felton, M., Garcia-Mila, M., & Gilabert, S. (2009). Deliberation versus dispute: The impact of argumentative discourse on learning and reasoning in the science classroom. *Informal Logic, 29*, 417–446.
Felton, M., Garcia-Mila, M., Villarroel, C., & Gilabert, S. (2015). Arguing collaboratively: Argumentative discourse types and their potential for knowledge building. *British Journal of Educational Psychology, 85*, 372–386.
Firman, M., Ertikanto, C., & Abdurrahman, A. (2019). Description of meta-analysis of inquiry-based learning of science in improving students' inquiry skills. *International Conference on Mathematics and Science Education*, 1–6 (Journal of Physics Conference Series).

Giamellaro, M. (2014). Primary contextualization of science through immersion in content-rich settings. *International Journal of Science Education, 36*, 2848–2871.

Gillies, R. M. (2007). *Cooperative learning: Integrating theory and practice*. Sage.

Gillies, R. M. (2009). *Evidence-based teaching: Strategies that promote learning*. Sense Publishers.

Gillies, R. (2016). Dialogic interactions in the cooperative classroom. *International Journal of Educational Research, 76*, 178–189.

Gillies, R., & Nichols, K. (2015). How to support primary teachers' implementation of inquiry: Teachers' reflections on teaching cooperative inquiry-based science. *Research in Science Education, 45*, 171–191.

Gillies, R., Nichols, K., Burgh, G., & Haynes, M. (2012). The effects of two strategic and meta-cognitive questioning approaches on children's explanatory behaviour, problem-solving, and learning during cooperative, inquiry-based science. *International Journal of Educational Research, 53*, 93–106.

Gillies, R., Nichols, K., Burgh, G., & Haynes, M. (2014). Primary students scientific reasoning and discourse during cooperative inquiry-based science activities. *International Journal of Educational Research, 63*, 127–140.

Howe, C., & Abedin, M. (2013). Classroom dialogue: A systematic review across four decades of research. *Cambridge Journal of Education, 43*, 325–356.

Huff, K., & Bybee, R. (2013). The practice of critical discourse in science classrooms. *Science Scope, 36*(9), 30–34.

Johnson, D., & Johnson, F. (2009). *Joining together: Group theory and group skills* (10th ed.). Allyn and Bacon.

Johnson, D., & Johnson, R. (2002). Learning together and alone: Overview and meta-analysis. *Asia Pacific Journal of Education, 22*, 95–105.

Johnson, D., Johnson, R., Roseth, C., & Shin, T. (2014). The relationship between motivation and achievement in interdependent situations. *Journal of Applied Social Psychology, 44*, 622–633.

Kang, J., & Keinonen, T. (2018). The effect of student-centered approaches on students' interest and achievement in science: Relevant topic-based, open and guided inquiry-based, and discussion-based approaches. *Research in Science Education, 48*, 865–885.

King, A. (1997). ASK to THINK-TEL WHY: A model of transactive peer tutoring for scaffolding higher level complex learning. *Educational Psychologist, 32*, 221–235.

Krajcik, J., & Sutherland, L. (2010). Supporting students in developing literacy in science. *Science, 328*, 456–459.

Llewellyn, D. (2014). *Inquire within: Implementing inquiry and argument-based science standards in grades 3–8* (3rd ed.). Corwin.

Murphy, C., Smith, G., & Broderick, N. (2019). A starting point: Provide children opportunities to engage with scientific inquiry and nature of science. *Research in Science Education, 51*(6), 1759–1793.

National Science Teachers Association (NSTA). (2004). *NSTA Position statement: Scientific inquiry*. NSTA.

Rennie, L. (2005). Science awareness and scientific literacy. *Teaching Science, 51*(1), 10–14.

Resnick, L., Michaels, S., & O'Connor, C. (2010). How (well structured) talk builds the mind. In D. Pressis & R. Sternberg (Eds.), *Innovations in educational psychology: Perspectives on learning, teaching and human development* (pp. 163–194). Springer.

Soysal, Y. (2021). Argument-based inquiry, teachers' talk moves, and students' critical thinking in the classroom. *Science & Education, 30*, 33–65.

Tang, K., & Moje, E. (2010). Relating multimodal representations to the literacies of science. *Research in Science Education, 40*, 81–85.

Treagust, D., Qureshi, S., Vishnumolakala, V., Ojeil, J., Mocerino, M., & Southam, D. (2020). Process-orientated guided learning inquiry (POGIL) as a culturally relevant pedagogy (CRP) in Qatar: A perspective from Grade 10 chemistry classes. *Research in Science Education, 50*, 813–831.

Tseng, C., Tuan, H., & Chin, C. (2013). How to help teachers develop inquiry teaching: Perspectives from experienced science teachers. *Research in Science Education, 43*, 809–825.

Chapter 9

Science as practice?

Jonathan Osborne

Introduction

Why practice? After all, most school science is trying to build a body of knowledge and understanding, so what has practice got to do with that? The argument here is that just knowing what something is simply not enough; rather, understanding depends on knowing how it came to be and – why it matters. And understanding how our current knowledge came to be requires some appreciation of the procedures and practices that scientists have used to derive that knowledge.

Ever since its inception, there has been a debate about what the function and purpose of science education is (Layton, 1973). How does science education achieve its goal of contributing to a liberal education? In the case of science, young people are the heirs to an inheritance of human achievements – beliefs, ideas, procedures, tools and artefacts that are highly valued (Fuller, 1989). However, society also has a need to train the next generation of scientists to sustain its economic well-being. And this has been the imperative that has dominated the goals of science education ever since the First World War (Committee to Enquire into the Position of Natural Science in the Educational System of Great Britain, 1918; Dainton, 1968; Lord Sainsbury of Turville, 2007; National Academy of Sciences, 2005; The Royal Society, 2014). The fundamental flaw with this as the major goal of science education is that it requires an emphasis on foundational knowledge – the bricks of the scientific edifice. And when approached in this manner, most students never get to see the edifice that constitutes their inheritance. If you want evidence of this, try asking any of your adult peers what they remember of value from their school science education. The hesitation and often silence speak volumes. In this chapter, I set out to offer a different vision of science education – one that would enable students to gain something of greater value – one that puts what you can do with scientific knowledge as the pre-eminent outcome.

So how does engaging in scientific practices help? First, just as attempting to play the violin makes you aware of the enormous level of skill and competence that the professional player has achieved, attempting to do science – to use the everyday practices of scientists – gives you a sense not only of the enormous

DOI: 10.4324/9781003137894-11

intellectual achievement scientific knowledge represents but also the epistemic foundations of how this knowledge is produced. In short, what do scientists *have to do* to establish reliable knowledge (Ziman, 1979) – knowledge that is 'true enough'(Elgin, 2017)?

Science as a practice-based activity

The emphasis on practices has emerged from the work of the science historians, philosophers, cognitive scientists and sociologists over the past 40 years. This view began with Kuhn's transformative work *The Structure of Scientific Revolutions* (Kuhn, 1962). The body of scholarship that followed has illuminated how science is actually done, both in the short-term (e.g., studies of activity in a particular laboratory or a program of study) and historically (e.g., studies of laboratory notebooks, published texts, eyewitness accounts) (Collins & Pinch, 1993; Conant, 1957; Geison, 1995; Latour & Woolgar, 1986; Pickering, 1995; Traweek, 1988). Seeing science as a set of practices has shown that theory development, reasoning and testing are elements of a larger ensemble of activities that include networks of participants and institutions (Latour, 1999; Longino, 2002); specialized ways of talking and writing (Bazerman, 1988); modelling with either mechanical and mathematical models or computer-based simulations (Nercessian, 2008); making predictive inferences; constructing appropriate instrumentation; and developing representations of phenomena (Latour, 1990; Lehrer & Schauble, 2006b).

Scientific practices are required to answer three questions about the material world (Ogborn, 1988; Osborne, 2011) that the sciences seek to answer. These are:

1   What exists? Which philosophers term the ontic question.
2   What causes this event/phenomenon to happen? Which is the causal question.
3   How do we know? Which is the epistemic question.

The primary goal is answering question 2 – to build explanatory models of what we see happening around us (Elgin, 2017; Lehrer & Schauble, 2006a). If "[m] odern science is one of humanities' greatest achievements" (Elgin, 2017, p. 1), then it behoves the science teacher to explain to their students why that is so and to at least provide a glimpse of how that has been achieved. Moreover, there is a need to be honest about what we are achieving in science – that we are building models and representations that affords an understanding of the phenomena, not one that replicates the phenomena itself. The Bohr model of the atom is not an accurate model of the atom; there is much more to inheritance and the expression of phenotypes than the insights offered by Mendelian genetics; the ideal gas laws are what they say they are – ideal. As Elgin (2017) points out, "[f]ar from being defects, models, idealizations, thought experiments figure ineliminably in successful science". To present them as some true account of the world, however, is to misrepresent their nature and to be intellectually dishonest with our students. How then are these explanatory models built? What practices are needed?

Asking questions

Questions are the engine which drives scientific research's need for explanation. As Asimov said, the most profound statement in science is not 'Eureka' but rather 'That's funny . . .' Likewise, only by pondering their own questions about the world can students begin to understand the importance of their role in science. The value of students' questions for learning has been emphasized by several authors (e.g., Biddulph et al., 1986; Chin & Osborne, 2008; Fisher, 1990; Penick et al., 1996). Questions raised by students activate their prior knowledge, focus their learning efforts, and help them elaborate on their knowledge (Schmidt, 1993). The act of 'composing questions' helps students to attend to the main ideas and check if the content is understood (Rosenshine et al., 1996). More importantly, knowing what the question is gives meaning to science. Textbooks, for instance, are full of explanations, but rarely begin by explicating the question that they are seeking to answer (D.J. Ford, 2006). As one ascerbic student commented, 'The problem with science is that it gives answers to questions you have never asked'.

Answering questions, however, requires students to engage in certain key practices. First, to answer the question of why something happens, there is an a priori requirement to establish what exists in the world. That requires a process of categorization and classification, as until something has been identified and defined, we cannot talk about it (Bowker & Leigh Star, 1999). This applies as much to concepts such as heat and temperature, speed and velocity, and chemical and physical change as it does to the attributes of species, the difference between a fish and a mammal and more. Making the distinction between heat and temperature, for instance, was one of the great intellectual achievements of the 18th century. Categorization and classification are still fundamental ongoing work in science. To date, over a million and a half species have been scientifically described, and there are still 5.5 million more to go (Pinker, 2018). Categorization and classification, however, require scientists to engage in the second practice of collecting, analyzing and interpreting data.

Analyzing and interpreting data

Observation and empirical enquiry produce data. To collect the data itself, however, there are many issues to be decided. First, what is to be measured, then how is it to be measured, then how much data to collect. The data may suffer random errors, and then some values may be outliers and should be eliminated from any data set. Second, there may be systematic errors in the data set. Moreover, data do not wear their meaning on their sleeves. They must be analyzed to identify what patterns can be found, and random patterns must be distinguished from distinct trends. This is fundamentally the work of the epidemiologist, ever since John Snow identified that there was a pattern to the people who were being infected with cholera coming to him, or the finding that skin cancers vary inversely with latitude (Osborne & Young, 1998). Such patterns lead to the search for a causal explanation and their aetiology. Epidemiology's greatest success perhaps was establishing the link between lung cancer and smoking (Fine et al., 2014).

Experiencing the cognitive effort that has to go into making sense of data is vital. For instance, in many Year 7 science classes you can observe students measuring the boiling point of water. But to what end? As an exercise to ascertain a value that has already been determined much more accurately by others, its purpose is highly questionable (Collins & Pinch, 1993). As an activity to develop students' facility with a thermometer, it has little value given that it takes little skill to read a digital thermometer. However, given that students' readings may vary considerably, the much more interesting question is how can we resolve the uncertainty? What methods exist and which of these are appropriate in this context? Since there are plural answers to the latter question, its resolution requires the application of a body of domain-specific, procedural knowledge to justify the most appropriate result. There must be room in the teaching of science for data sets that are ambiguous and where the meaning is less than self-evident. This is why the software tools offered by CODAP (https://codap.concord.org/) and Tuva Datalabs (https://tuvalabs.com/) are such a valuable means of exposing students to data sets where the pattern is not linear and the explanation of the pattern is not self-evident.

Moreover, the PISA assessment framework for the 2025 Assessment sees the competency to "Evaluate designs for scientific enquiry and interpret scientific data and evidence critically" as one of the three core competencies of a scientifically literate person. Such a competency can only develop by asking students either to gather their own sets of data or use secondary data sets, and then establish and justify the best interpretation.

Developing and using models

Answering the causal question of why it happens requires a hypothesis. For instance, the question of why you look like your parents obviously leads to an understanding that something is passed on at the point of conception. But how is it passed on? In what way is that information encoded? Building an understanding of how invariably requires the construction of a model – in this case, a model of the DNA structure (Watson & Crick, 1953).

Models are needed in the sciences because the sciences deal with things too large to imagine or too small to see, such as a cell, the inside of the human body, or the atom itself (Harrison & Treagust, 2002) (Gilbert & Boulter, 2000). Young children will begin by constructing simple physical models or diagrammatic representations (Ainsworth et al., 2011) but then, at higher grades, models become more abstract and more reliant on mathematics. However, although some models, such as the Bohr model of the atom or a model of a cell, are simply representational, other models can explain or enable predictions. Thus, the bicycle chain model of an electric circuit not only explains why the electric light comes on instantaneously but also explains why, if the chain is broken, no light will come on. Nobody imagines that electrons are connected like links in a bicycle chain. Models are "felicitous falsehoods" (Elgin, 2017). While they may not be true, they help us build understanding and, as such, are epistemically valuable. As Lehrer and Schauble argue, "[m]odeling is a form of disciplinary argument" that students must "learn to participate in over a long and extended period of practice" (p. 182, Lehrer & Schauble, 2006b). In a very real sense, the activity of modelling and discussing the extent to which they are true lays bare the soul of science.

Constructing explanations

Models are essential for explanations. Whether it is the particle model of solid, liquids and gases, the Bohr model of the atom or a model of animal cell, once in possession of a model it can be used to construct an explanation. For instance, the particle model of a gas explains how ammonia on one side of the room can travel to the other side in a sealed room. The construction of an explanation, however, requires cognitive effort (Chi, 2009; Hatano & Inagaki, 1991) – cognitive effort that helps to develop understanding, as any teacher will recognize. As a teacher, the first time you are forced to explain something to a class requires considerable thought if you are to do it well. And what is good for the teacher is good for his or her students. Chi et al. (2017) postulate the reason that asking students to engage in explanation is effective is that articulating an incorrect explanation produces conflict. The outcome of such conflict will require reflection and self-repair by the student if resolution is to occur. Commonly, however, students are the recipients of many explanations provided by teachers, but rarely are they asked to construct explanations themselves (McRobbie & Thomas, 2001; Weiss et al., 2003). Given its value, one must ask why?

Arguing from evidence

Explanations must be argued for, however. They must be coherent with the data, be capable of explaining phenomena and be capable of making predictions. Argument, then, is a central feature of science, as the construction of knowledge is a dialectic between construction and critique (Ford, 2008). The model in Figure 9.1, which has emerged from studies of what scientists do, shows the centrality of argument to the sciences (Klahr et al., 1993). Klahr et al. (1993) characterize science as taking place in three spaces. On the right, the main activity is hypothesizing in an attempt to develop explanations for what has emerged from activity taking place in the investigation space involving experimentation,

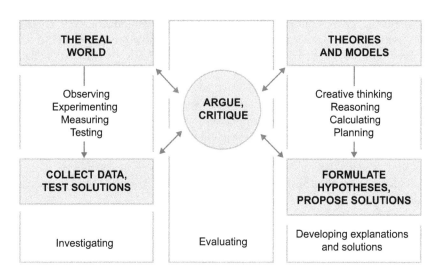

THE REAL
WORLD

Observing
Experimenting
Measuring
Testing

ARGUE,
CRITIQUE

THEORIES
AND MODELS

Creative thinking
Reasoning
Calculating
Planning

COLLECT DATA,
TEST SOLUTIONS

Investigating

Evaluating

FORMULATE
HYPOTHESES,
PROPOSE SOLUTIONS

Developing explanations
and solutions

Figure 9.1 A model of scientific activity

represented on the left. The third sphere of activity, the central one, is evaluating whether the explanation fits with the evidence or whether the evidence is satisfactory.

These spheres of activity are not self-evident, however. That is, someone coming recently to science would not necessarily recognize these three problem spaces as the key features of the work of the scientist. Studies of young children reveal that students often see science differently, suggesting that many students adopt a naive interpretation of scientific theories, seeing them as emerging from data (Driver et al., 1996; Klahr & Carver, 1995; Schauble et al., 1991). The outcome is that too many students fail to see that it is really theories that are the "crowning glory of science" (Harré, 1984); that science is fundamentally about *ideas*, not experimentation; and that your name is preserved for posterity in science not for devising a new experiment, but for developing a new theory. Argument, then, is a central and core feature of the deliberative process of scientific ideas. Indeed, the history of science is better seen as a history of mistakes resolved by argument and the gradual emergence of consensus (Allchin, 2012).

Argument is also central to the activity of learning science. Why? Because scientific ideas are hard to understand. This is not simply because they require coming to terms with an ontological zoo of unfamiliar entities – ions, electrons, atoms, cells, chromosomes, genes, molecules, quanta, radiation, nuclei, electric charge, voltage, current, wavelength, frequency and more – but because the big ideas of science are fundamentally unnatural (Cromer, 1993; Wolpert, 1992). From the idea that day and night are caused by a spinning Earth rather than a moving Sun, to the idea that you look like your parents because every cell in your body carries a chemically coded blueprint about how to reproduce yourself, science challenges our common sense. The teacher of science is not just a teacher of a foreign language, but also teacher of a set of what to many would seem to be "crazy ideas" (Osborne, 2019). Table 9.1 shows one list of some of these 'crazy' ideas.

Yes, these can be taught as 'facts', but the evidence is that such a process is remarkably ineffective (Chi, 2009; Hake, 1998; Osborne, 2010). Rather, understanding develops when students are offered an opportunity to engage in deliberative argumentation, examine the ideas they hold and compare their thinking to what is being offered by their teacher of science. In essence, what any good educator does is engage in a conversation which starts by asking the students to 'tell me how you explain this phenomenon' and 'I will tell you how and why I see it my way'. Research shows that starting with an exploration of student conceptions leads to more effective outcomes (Sadler et al., 2013), and there is a considerable body of evidence that shows that student learning is more effective when students are given an opportunity to engage in deliberative discussions and argumentation (Alexander, 2020; Chi, 2009; Chi et al., 2017; Mercer et al., 2004; Osborne, 2010; Smith et al., 2009; Zohar & Nemet, 2002).

Opportunities for argumentation can be provided using concept maps, concept cartoons, the 'turn-and-talk' strategy where students talk to their neighbour, argument lines, four corners and computer-based platforms such as Kahoot, Socrative and Brain Candy. The advantage of many of these approaches is that they sustain a separation between the idea and the student. That is, the student

Table 9.1 Ideas which are unnatural, uncommon sense or that would appear 'crazy' when first proposed

1	The Earth is a sphere, not flat
2	Airplanes are supported by the air
3	Day and night are caused by a spinning Earth
4	The continents have moved
5	We have evolved from other animals
6	The Earth is 5 billion years old
7	Diseases are caused by tiny living organisms
8	The universe started with a big bang
9	Time slows down when you travel at the speed of light
10	Most of the atom is empty space
11	Heavier things do not always fall faster
12	Most of the atoms that make up our bodies were created in stars a long time ago
13	Human organisms are a way of ensuring the survival of DNA
14	There are more bacteria in the human mouth than there are people on the planet
15	Plants can produce complex molecules in a way that we humans cannot replicate
16	All substances are made from only 80 different elements
17	Most of the matter in a plant comes from the air

does not own the idea and does not run the risk of the embarrassment of being shown to be wrong. Whole-class discussion and argumentation is a more sophisticated skill which requires careful orchestration, establishing classroom norms and support with the use of open-ended questions and appropriate talk moves (Michaels & O'Connor, 2012; Osborne et al., 2016).

There is another important reason for giving students the opportunity to engage in argumentation. Argumentation requires discourse, and discourse means using the language of science. Given that learning science is fundamentally about learning a new language (Norris & Phillips, 2003), and that "every lesson is a language lesson" (Wellington & Osborne, 2001, p. 2), acquiring confidence with a new language, as any teacher of languages will tell you, is dependent on opportunities to practice and use that language (van Lier, 1996). As Lemke points out, "The one single change in science teaching that should do more than any other to improve students' ability to use the language of science is to *give them more practice actually using it*"[1] (p. 168).

The dominant form of talk in most classrooms, but particularly science classrooms, is triadic dialogue (Lemke, 1990), which others call initiation-response-evaluation (IRE). The features of this dialogue are that it is initiated by the teacher with a question. The student commonly offers a short, phrase-like response, and then the teacher provides feedback, which is commonly of an evaluative nature: e.g., teacher: 'What is the symbol for Calcium?', student: 'Ca', teacher: 'Correct'. This form of linguistic practice has three problems. First, this dialogue is unnatural. Normally the person who does not know (the student) is the person who asks the question, rather than the teacher who does know. Second, it requires negligible use of language by the student, minimizing opportunities to practice the language of science.

Why critique is important to scientific learning, then, can be understood by considering the process of learning as being metaphorically akin to weighing the two competing ideas – proposition A and proposition B. Initially, evidence might suggest they are both of equal merit. Imagine then that a good explanation, in the absence of critique, raises the credibility of A, such that the weight of evidence for this proposition is now twice as strong as the evidence for idea B. But when idea B is critiqued, its credibility is halved as well. Now the ratio of credible evidence for A with respect to B becomes four times as strong, leading to a more secure and enduring understanding (Ames & Murray, 1982; Hynd & Alvermann, 1986; Schwarz et al., 2000).

Why, then, the virtual absence of critique and critical thinking from science education, both past and present? This is an absence which at least explains Rogers' comment that "[w]e should not assume that mere contact with science, which is so critical, will make students think critically" (Rogers, 1948, p. 7). The pretense that science education makes its students critical and analytical thinkers is one of the strongest arguments for engaging students in the practice of argumentation as it places the higher-order skills of critique and evaluation at the center of teaching and learning science, requiring students to engage in evaluation, synthesis and comparison and contrast (Ford & Wargo, 2011). Another argument for argument is that there are a range of types of arguments in science, and students need to experience these. There are arguments which are abductive[2] (inferences to the best possible explanation), such as Darwin's arguments for the theory of evolution; hypothetico-deductive, such as Pasteur's predictions about the outcome of the first test of his anthrax vaccine; or simply inductive generalizations archetypal represented by 'laws'.

Obtaining, evaluating and communicating information

This practice is core to all the sciences. Literacy is not some kind of adjunct to science – it is *constitutive of science itself* (Norris & Phillips, 2003). Indeed, contrary to the popular image which perceives the major practice of science to be one of 'doing experiments', Tenopir and King (2004) found that engineers and scientists spend more than 50% of their time engaged in reading and writing science. In short, reading, writing and arguing are core activities for *doing* science (Lemke, 1990; Jetton & Shanahan, 2012). And, as Norris and Phillips (2003) point out, the fundamental sense of scientific literacy is the ability of an individual to use the language of science to communicate with other scientists. The five major communicative activities of science can be seen as writing science, talking science, reading science, 'doing'[3] science and representing scientific ideas – an idea which is represented by Figure 9.2.

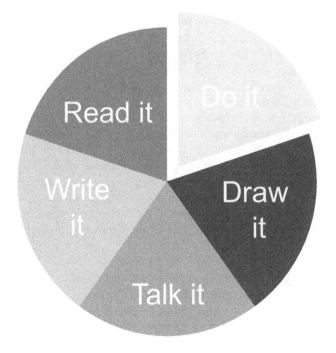

Figure 9.2 The major activities of science

So to develop some kind of insight into the activity of science, science education must offer students the opportunity to experience and practice a broad range of discursive and literate activities and help them build the disciplinary literacy required (Pearson et al., 2010; Shanahan & Shanahan, 2008). As the work of systemic functional linguists has pointed out, the forms of communication within a discipline, while reliant on basic literacy, are often specific *to* the discipline and that there is such a thing as 'disciplinary literacy' – that is, the ability to decode and interpret more complex forms of text, and to recognize the nature and function of genre specific to the discipline (Fang, 2006; Halliday & Martin, 1993; Lemke, 1990; Martin & Veel, 1998; Schleppegrell & Fang, 2008).

Another major challenge posed by the language of science, however, is the use of 'academic language' – the form of language that is used to attain "conciseness, achieved by avoiding redundancy; using a high density of information-bearing words, ensuring precision of expression; and relying on grammatical processes to compress complex ideas into few words" (Snow (2010, p. 450). Yet, traditionally, science teachers have not paid much attention to text (Wellington & Osborne, 2001), operating rather on the notion that there is nothing particularly distinctive about the genres in which science is communicated. The outcome of the view that literacy is seen as a peripheral feature of science results in science teachers who lack a knowledge about "the vital role [that] literacy plays in enhancing rather than replacing science learning" (Pearson et al., 2010, p. 462) and thus teachers fail to mentor students in the necessary literate practices (Barton et al., 2002), which would help them read and write science. Wellington and Osborne (2001) provide a range of ideas on how to support writing and reading science, as does the website Reading to Learn in Science (www.serpinstitute.org/reading-science). Support for talking in science can be found in the *Talk Science Primer* (Michaels & O'Connor, 2012) and in *Arguing From Evidence in Middle School Science: 24 Activities for Productive Talk and Deeper Learning* (Osborne et al., 2016).

There is another reason this practice is so important. Across the globe, there is a concern about the ease with which people accept beliefs for which there is no material evidence, and for which there is often good evidence to the contrary. Whether it is the idea that the Earth is flat, that vaccines cause autism or that climate change is a hoax, the willingness of individuals to accept irrational beliefs – irrational in the sense that there is no evidence to support them – is of grave concern (Kozyreva et al., 2020; Lee et al., 2021). Some of these beliefs – for instance, that vaccines are harmful – endanger not only the lives of those who hold them, but the whole community that depends on a high level of vaccination to ensure its health. Yet:

> the rationality of science is secured by its commitment to evidence; the fostering of a commitment to evidence is a fundamental educational aim . . . and science education can, and should be seen as a central component of an education dedicated to the fostering of rationality and critical thinking.
>
> (Siegel, 1989, p. 11)

Underlying the growing concern about the failings of science education to develop this ability is the use of social media to present and circulate such false beliefs. The essential problem is that traditional sources of information have been displaced by unregulated open access through both the Internet, which bypasses experts and certified authorities, and social media, where misinformation can spread rapidly and widely through existing social networks (Vosoughi et al., 2018). Without a traditional professional gatekeeper, how then does the individual sort truth from falsehood or deceptive half-truth? This is not some form of cross-disciplinary competence, but has specific disciplinary elements (Caulfield, 2016). Its importance is such that the competency to 'Research, evaluate and use scientific information for decision making' will be assessed in the OECD PISA tests in 2025.

Essential to this competency is an understanding of how trustworthy scientific knowledge is produced and "how monied interests try to 'bend' science, present pseudoscience as science, portray reliable science as 'junk science,' or foster an image of uncertainty even where scientific experts have reached a solid consensus" (Höttecke & Allchin, 2020, p. 4). The scientifically educated individual now needs a basic understanding of the transformations that occur in the trajectory from 'test tube to YouTube' or from 'lab books to Facebook'. Second, an educated person would know that there is a tendency of all individuals to exist in a 'bubble' supported by news that reinforces the views and perceptions they already have. Hence, all scientific information should be approached with a policy of circumspection which seeks to determine if there are contradictory findings, questions the credibility of the source and recognizes the limits and potential prejudices of the channel of communication in reporting the findings. For instance, has the finding been reported by somebody with professional expertise in the relevant domain of science?

At the core of this competency is an understanding that science is a communal enterprise and that science is not infallible. In many ways, error is common in science (Allchin, 2012). However, while individual scientists or teams might be mistaken, the community is more trustworthy, and the goal of the sciences is closure, which is attained when there is overwhelming consensus. Moreover, that consensus is a consensus of experts (Crease & Selinger, 2006; Oreskes, 2019). Nor is science a democracy where one scientist's views are as valuable as another's. The views that matter are from those who have relevant background knowledge, skills in interpreting particular results and awareness of potential flaws in reasoning. Thus, a petition claiming that vaccines are dangerous is meaningless if the signatories are not scientific experts on vaccination. However, research shows that students are really poor at this competence (Breakstone et al., 2021). Students, therefore, need to be educated with the generic epistemic tools that might enhance their critical facility (Wineburg & McGrew, 2019), and with the epistemic knowledge that they need. Given that so many of these issues have a scientific component, it is *essential* that this capability is developed and supported by science education.

Implications and conclusions

Across the globe, people are expecting and demanding more of their education systems. Developing knowledge and understanding is not seen as sufficient. Rather, the focus today is on developing competency – that is, the ability to use the knowledge to do something with it and, in essence, to demonstrate a level of proficiency at certain practices (Baker, 2014; Koeppen et al., 2008; OECD, 2016; Rychen & Salganik, 2003). The primary purpose of engaging in practice is to develop students' facility with that practice, how that practice contributes what we know, and how that practice helps to build reliable knowledge. Knowledge of how we know in the science (knowing how) is reliant on a developing a body of procedural knowledge or concepts of evidence (Gott et al., 2008). Knowing why such practices are necessary is dependent on epistemic knowledge. Such disciplinary knowledge, I would argue, is not some superficial adjunct, but is core to any science education that claims to build an understanding of the sciences and their value. While there are clearly more practices that students will engage in learning science, such as 'Planning and carrying out investigations' and 'Using mathematical and computational thinking' and more, I would argue that those discussed here are the most important. Few individuals will need to plan and conduct scientific inquiry in their lives – all of them will need to evaluate information they find on the Internet.

Questions for further debate

1   What would an emphasis on practices mean for the teaching of science compared to what you do now?
2   Which practice is most readily taught, and which would be the most challenging and why?
3   How well has teachers' own education prepared them to teach these practices – for instance, argument in science?

Suggested further reading

National Research Council. (2012). *A framework for K-12 science education: Practices, crosscutting concepts, and core ideas.* https://nap.nationalacademies.org/catalog/13165/a-framework-for-k-12-science-education-practices-crosscutting-concepts

Osborne, J. F. (2014). Teaching scientific practices: Meeting the challenge of change. *Journal of Science Teacher Education, 25,* 177–196.

Notes

1 Emphasis added.
2 Abductive arguments are also known as retroductive arguments.
3 In one sense, any of these activities could be said to be 'doing science'. In this chapter, the term 'doing science' is used to refer to the act of engaging in empirical inquiry.

References

Ainsworth, S., Prain, V., & Tytler, R. (2011). Drawing to learn in science. *Science, 333*, 1096–1097.

Alexander, R. (2020). *A dialogic teaching companion*. Routledge.

Allchin, D. (2012). Teaching the nature of science through scientific errors. *Science Education, 96*(5), 904–926. https://doi.org/10.1002/sce.21019

Ames, G. J., & Murray, F. B. (1982). When two wrongs make a right: Promoting cognitive change by social conflict. *Developmental Psychology, 18*, 894–897.

Baker, D. (2014). *The schooled society: The educational transformation of global culture*. Stanford University Press.

Barton, M. L., Heidema, C., & Jordan, D. (2002). Teaching in mathematics, reading, and science. *Educational Leadership*, 24–28.

Bazerman, C. (1988). *Shaping written knowledge*. The University of Wisconsin Press.

Biddulph, F., Symington, D., & Osborne, R. (1986). The place of children's questions in primary science education. *Research in Science and Technological Education, 4*, 77–88.

Bowker, G. C., & Leigh Star, S. (1999). *Sorting things out: Classification and its consequences*. MIT Press.

Breakstone, J., Smith, M., Wineburg, S., Rapaport, A., Carle, J., Garland, M., & Saavedra, A. (2021). *Students' civic online reasoning: A national portrait*. Educational Researcher.

Caulfield, M. (2016). *Yes, digital literacy. But which one?* https://hapgood. us/2016/12/19/yes-digital-literacy-but-which-one/

Chi, M. (2009). Active-constructive-interactive: A conceptual framework for differentiating learning activities. *Topics in Cognitive Science, 1*, 73–105.

Chi, M., Kang, S., & Yaghmourian, D. L. (2017). Why students learn more from dialogue-than monologue-videos: Analyses of peer interactions. *Journal of the Learning Sciences, 26*(1), 10–50. https://doi.org/10.1080/10508406.2016.1204546

Chin, C., & Osborne, J. (2008). Students' questions: A potential resource for teaching and learning science. *Studies in Science Education, 44*(1), 1–39.

Collins, H., & Pinch, T. (1993). *The Golem: what everyone should know about science*. Cambridge University Press.

Committee to Enquire into the Position of Natural Science in the Educational System of Great Britain. (1918). *Report of the committee appointed by the Prime Minister to enquire into the position of natural science in the educational system of Great Britain*. https://archive.org/details/reportofcommitte00grearich/page/n3/mode/2up?ref=ol&view=theater

Conant, J. (1957). *Harvard case histories in experimental science* (Vol. 1 & 2). Harvard University Press.

Crease, R. P., & Selinger, E. (2006). *The philosophy of expertise*. Columbia University Press.

Cromer, A. (1993). *Uncommon sense: The heretical nature of science*. Oxford University Press.

Dainton, F. S. (1968). *The Dainton report: An inquiry into the flow of candidates into science and technology*. HMSO.

Driver, R., Leach, J., Millar, R., & Scott, P. (1996). *Young people's images of science*. Open University Press.

Elgin, C. Z. (2017). *True enough*. Harvard University Press.

Fang, Z. (2006). The language demands of science reading in middle school. *International Journal of Science Education, 28*(5), 491–520.

Fine, P. E. M., Goldacre, B., & Haines, A. (2014). Epidemiology – A science for the people. *The Lancet, 381*, 1249–1252.

Fisher, R. (1990). *Teaching children to think*. Simon and Schuster.

Ford, D. J. (2006). Representations of science within children's trade books. *Journal of Research in Science Teaching, 43*(2), 214–235.

Ford, M. J. (2008). Disciplinary authority and accountability in scientific practice and learning. *Science Education, 92*(3), 404–423. http://dx.doi.org/10.1002/sce.20263

Ford, M. J., & Wargo, B. M. (2011). Dialogic framing of scientific content for conceptual and epistemic understanding. *Science Education, 96*(3), 369–391.

Fuller, T. (Ed.) (1989). *The voice of liberal learning: Michael Oakshott on education*. Yale University Press.

Geison, J. (1995). *The private science of Louis Pasteur*. Princeton University Press.

Gilbert, J., & Boulter, C. (Eds.). (2000). *Developing models in science education*. Kluwer.

Gott, R., Duggan, S., & Roberts, R. (2008). *Concepts of evidence*. School of Education, University of Durham.

Hake, R. R. (1998). Interactive-engagement versus traditional methods: A six-thousand-student survey of mechanics test data for introductory physics courses. *American Journal of Physics, 66*(1), 64–74. http://link.aip.org/link/?AJP/66/64/1

Halliday, M. A. K., & Martin, J. R. (1993). *Writing science: Literacy and discursive power*. Falmer Press.

Harré, R. (1984). *The philosophies of science: An introductory survey* (2nd ed.). Oxford University Press.

Harrison, A. G., & Treagust, D. F. (2002). A typology of school science models. *International Journal of Science Education, 22*(9), 1011–1026.

Hatano, G., & Inagaki, K. (1991). Sharing cognition through collective comprehension activity. In L. Resnick, J. M. Levine, & S. D. Teasley (Eds.), *Perspectives on socially shared cognition* (pp. 331–348). American Psychological Association.

Höttecke, D., & Allchin, D. (2020). Reconceptualizing nature-of-science education in the age of social media. Science Education, 104(4), 641–666. https://doi.org/10.1002/sce.21575

Hynd, C., & Alvermann, D. E. (1986). The role of refutation text in overcoming difficulty with science concepts. *Journal of Reading, 29*(5), 440–446.

Jetton, T. L., & Shanahan, C. (Eds.). (2012). *Adolescent literacy in the academic disciplines: General principles and practical strategies*. Guilford Press.

Klahr, D., & Carver, S. M. (1995). Scientific thinking about scientific thinking. *Monographs of the Society for Research in Child Development, 60*(4), 137–151.

Klahr, D., Fay, A. L., & Dunbar, K. (1993). Heuristics for scientific experimentation: A developmental study. *Cognitive Psychology, 24*(1), 111–146.

Koeppen, K., Hartig, J., Klieme, E., & Leutner, D. (2008). Current issues in competence modeling and assessment. *Journal of Psychology, 216*(2), 61–73.

Kozyreva, A., Lewandowsky, S., & Hertwig, R. (2020). Citizens versus the internet: Confronting digital challenges with cognitive tools. *Psychological Science in the Public Interest, 21*(3), 103–156. https://doi.org/10.1177/1529100620946707

Kuhn, T. E. (1962). *The structure of scientific revolutions.* University of Chicago Press.

Latour, B. (1990). Visualisation and cognition: Drawing things together. *Representation in Scientific Activity,* 19–68.

Latour, B. (1999). *Pandora's hope: Essays on the reality of science studies.* Harvard University Press.

Latour, B., & Woolgar, S. (1986). *Laboratory life: The construction of scientific facts* (2nd ed.). Princeton University Press.

Layton, D. (1973). *Science for the people: The origins of the school science curriculum in England.* Allen and Unwin.

Lee, C. D., White, G., & Dong, D. (2021). *Educating for civic reasoning and discourse.* National Academy of Education.

Lehrer, R., & Schauble, L. (2006a). Cultivating model-based reasoning in science education. In R. K. Sawyer (Ed.), *The Cambridge handbook of the learning sciences* (pp. 371–387). Cambridge University Press.

Lehrer, R., & Schauble, L. (2006b). Scientific thinking and science literacy. In W. Damon, R. M. Lerner, & N. Eisenberg (Eds.), *Handbook of child psychology* (pp. 153–196). Wiley.

Lemke, J. (1990). *Talking science: Language, learning and values.* Ablex Publishing.

Longino, H. E. (2002). *The fate of knowledge.* Princeton University Press.

Lord Sainsbury of Turville. (2007). *The race to the top: A review of government's science and innovation policies.* HMSO.

Martin, J. R., & Veel, R. (1998). *Reading science.* Routledge.

McRobbie, C., & Thomas, G. (2001). They don't teach us to explain, they only tell us other people's explanations. Paper presented at the European Association for Research on Learning, Freiburg.

Mercer, N., Dawes, L., Wegerif, R., & Sams, C. (2004). Reasoning as a scientist: ways of helping children to use language to learn science. *British Education Research Journal, 30*(3), 359–377.

Michaels, S., & O'Connor, C. (2012). *Talk science primer.* TERC.

National Academy of Sciences. (2005). *Rising above the gathering storm: Energizing and employing America for a brighter economic future.* National Academies Press.

Nercessian, N. (2008). Model-based reasoning in scientific practice. In R. A. Duschl & R. E. Grandy (Eds.), *Teaching scientific inquiry: Recommendations for research and implementation* (pp. 57–79). Sense.

Norris, S. P., & Phillips, L. (2003). How literacy in its fundamental sense is central to scientific literacy. *Science Education, 87,* 224–240.

OECD. (2016). *PISA 2015 assessment and analytical framework: Science, reading, mathematic and financial literacy.* PISA, OECD Publishing.

Ogborn, J. (1988). *A map of science personal submission to the National Curriculum Working Group on Science Education.* Ref. No. NC/SWG (88) EV, 65.

Oreskes, N. (2019). *Why trust science?* Princeton University Press.

Osborne, J. F. (2010). Arguing to learn in science: The role of collaborative, critical discourse. *Science, 328,* 463–466.

Osborne, J. F. (2011). Science teaching methods: A rationale for practices. *School Science Review, 93*(343), 93–103.

Osborne, J. F. (2019). Not "hands on" but "minds on": A response to Furtak and Penuel. *Science Education,* 103(5), 1280–1283. https://doi.org/10.1002/sce.21543

Osborne, J. F., Donovan, B., Henderson, B., MacPherson, A., & Wild, A. (2016). *Arguing from evidence in middle school science: 24 activities for productive talk and deeper learning*. Corwin Press.

Osborne, J. F., & Young, A. R. (1998). The biological effects of ultra-violet radiation: A model for contemporary science education. *Journal of Biological Education, 33*(1), 10–15.

Pearson, D., Moje, E. B., & Greenleaf, C. (2010). Literacy and science: Each in the service of the Other. *Science, 328*, 459–463.

Penick, J. E., Crow, L. W., & Bonnsteter, R. J. (1996). Questions are the answers. *Science Teacher, 63*, 26–29.

Pickering, A. (1995). *The mangle of practice: Time, agency & science*. University of Chicago Press.

Pinker, S. (2018). *Enlightenment NOW: The case for reason, science, humanism and progress*. Allen Lane.

Rogers, E. M. (1948). Science in general education. In E. J. McGrath (Ed.), *Science in general Education*. Wm.C, Brown Co.

Rosenshine, B., Meister, C., & Chapman, S. (1996). Teaching students to generate questions: A review of the intervention studies. *Review of Educational Research, 66*, 181–221.

Rychen, D. S., & Salganik, L. H. (Eds.). (2003). *Definition and selection of key competencies: Executive summary*. Hogrefe.

Sadler, P. M., Sonnert, G., Coyle, H. P., Cook-Smith, N., & Miller, J. L. (2013). The influence of teachers' knowledge on student learning in middle school physical science classrooms. *American Educational Research Journal, 50*(5), 1020–1049.

Schauble, L., Klopfer, L. E., & Raghavan, K. (1991). Students' transition from an engineering model to a science model of experimentation. *Journal of Research in Science Teaching, 28*(9), 859–882.

Schleppegrell, M., & Fang, Z. (2008). *Reading in secondary content areas: A language-based pedagogy*. University of Michigan Press.

Schmidt, H. G. (1993). Foundations of problem-based learning: Rationale and description. *Medical Education, 17*, 11–16.

Schwarz, B. B., Neuman, Y., & Biezuner, S. (2000). Two wrongs may make a right . . . if they argue together! *Cognition and Instruction, 18*(4), 461–494.

Shanahan, T., & Shanahan, C. (2008). Teaching disciplinary literacy to adolescents: Rethinking content area literacy. *Harvard Educational Review, 78*(1), 40–59.

Siegel, H. (1989). The rationality of science, critical thinking and science education. *Synthese, 80*(1), 9–42.

Smith, M. K., Wood, W. B., Adams, W. K., Wieman, C., Knight, J. K., Guild, N., & Su, T. T. (2009). Why peer discussion improves student performance on in-class concept questions. Science, 323(5910), 122–124. https://doi.org/10.1126/science.1165919

Snow, C. (2010). Academic language and the challenge of reading for learning about science. *Science, 328*, 450–452.

Tenopir, C., & King, D. W. (2004). *Communication patterns of engineers*. Wiley.

The Royal Society. (2014). *Vision for science and mathematics education*. The Royal Society Science Policy Centre. https://royalsociety.org/~/media/education/policy/vision/reports/vision-summary-report-20140625.pdf

Traweek, S. (1988). *Beamtimes and lifetimes: The world of high energy physicists.* Harvard University Press.

van Lier, L. (1996). *Interaction in the language curriculum.* Longman.

Vosoughi, S., Roy, D., & Aral, S. (2018). The spread of true and false news online. *Science, 359*(6380), 1146–1151. https://doi.org/10.1126/science.aap9559

Watson, J. D., & Crick, F. H. C. (1953). A structure for deoxyribose nucleic acid. *Nature, 171,* 737–738.

Weiss, I. R., Pasley, J. D., Sean Smith, P., Banilower, E. R., & Heck, D. J. (2003). *A study of K-12 mathematics and science education in the United States.* Horizon Research. http://www.horizon-research.com/insidetheclassroom/reports/looking/frontmatter.pdf

Wellington, J., & Osborne, J. F. (2001). *Language and literacy in science education.* Open University Press.

Wineburg, S., & McGrew, S. (2019). Lateral reading and the nature of expertise: Reading less and learning more when evaluating digital information. *Teachers College Record, 121*(11), n11.

Wolpert, L. (1992). *The unnatural nature of science.* Faber and Faber.

Ziman, J. (1979). *Reliable knowledge.* Cambridge University Press.

Zohar, A., & Nemet, F. (2002). Fostering students' knowledge and argumentation skills through dilemmas in human genetics. *Journal of Research in Science Teaching, 39*(1), 35–62.

Chapter 10

Learning and assessment

Erin Marie Furtak

Introduction

As a classroom teacher getting my start in the early 2000s, I was excited to engage my students in the process of science in my classroom. Like many other teachers at the schools where I worked, I created lessons for my students to experience in the process of scientific inquiry, or the skills and habits of mind of scientists, as they conducted their daily work.

In one of these investigations, students had a chance to make predictions about and measure the height of plastic poppers: small rubber pop toys that, when turned inside out and then placed on a surface, pop high up into the air. The idea of the lab was to engage students in designing their own investigations of the poppers, examining how setting it on different surfaces like a hard desk or a carpet might influence the height it popped.

This activity – similar to many (Chen & Klahr, 1999; Klahr & Nigam, 2004) – abstracts the process of science from the science concepts that are related to the height the popper reaches. That is, students did not need to rely on any prior knowledge of kinetic or potential energy, or elastic collisions, in order to engage in the activity. An assessment that could follow such an investigation would similarly focus on how students identified and controlled variables, collected measurements, and used evidence to support their conclusions (e.g., Shavelson et al., 1991).

Critics have questioned the value of teaching and assessing science inquiry in this way, and indeed have asserted that it does not actually lead to students engaging in the process of real science, which relies greatly on scientists' understanding of key ideas and concepts, and their ability to apply these understandings as they design experiments, conduct data analyses, develop hypotheses, revise models, and so on. That is, it does not rely upon the ways in which science practice is embedded in disciplines of science, and specific knowledge and understandings. (Ford, 2015; Ford & Forman, 2006).

While teaching the approaches of the 'scientific method' or 'science inquiry' in this way may have been done in the past, international reform efforts now prioritize the strategic application and use of key science concepts as part of science

DOI: 10.4324/9781003137894-12

learning, in effect intertwining or braiding together students' engagement in processes of inquiry with foundational scientific understandings in science (e.g., NRC, 2012, OECD, 2015). Assessments, in turn, should be designed in ways that provide students opportunities to respond to scenarios in ways that demonstrate both their grasp of the process of inquiry as well as their understandings of important science concepts (NRC, 2014; Rönnebeck et al., 2018).

This chapter will provide a quick review of the difference in – and the fundamental relationship between – content and process in science learning and assessment, with examples. Then, it will provide some frameworks for designing assessments that integrate both content and process.

Definitions and historical perspectives

Stemming back to debates in the mid-19th century with reformers such as Pestalozzi, science educators have grappled with a fundamental tension in teaching science. As Schwab (1962) described it, this tension exists between the process of inquiry, or the habits of mind and activities of scientists, as they engage in the pursuit of scientific knowledge. At the same time, there is the product of this process of scientific inquiry, which embodies the major concepts and laws – ever evolving – that are produced. Both are goals in science teaching, but they have not been equally prioritized.

Science content

A major element of science education has always been what Schwab (1962) called the 'product' of science knowledge, or the facts and concepts produced by scientists engaging in their daily work (Latour & Woolgar, 1979). Indeed, leaders and educational reformers emphasize that students should have the opportunity to learn these major ideas of science through participation in school. Standards documents commonly include lists of these ideas (e.g., AAAS, 1990), sorted according to the grades at which students may learn them, and increasing in sophistication and complexity as students advance in school; for example, students may learn that shadows differ in length and position in primary school prior to learning about the relative position of the Sun to the Earth and the Earth's orientation to the Sun at different times of day, and different seasons of the year.

When it comes to designing assessments, these kinds of concepts are sometimes considered more straightforward to assess. These smaller pieces can be broken down into fill-in-the-blank, multiple-choice questions that can be easily graded and used to determine what students know. These kinds of smaller bits of 'atomized' knowledge are also easier to reliably assess through multiple-choice or fill-in-the-blank items based on students' recall of facts (Shepard, 2000). However, these assessments can further be designed so that they provide more information than whether students get the main idea or not (Otero & Nathan, 2008). Instead, assessments can be intentionally designed to draw students' common ideas and experiences around these science concepts (e.g., that the Earth is closer to the Sun when it's winter in the Northern Hemisphere), thus providing teachers with diagnostic information about what students know and what they still need support to learn (e.g., Briggs et al., 2006; Sadler, 1998).

Science inquiry

At the same time, educators have also sought to assess the process of science, or the process of inquiry, for decades. These assessments take a different tactic than those that look at concepts, but rather seek to capture the ways in which students exhibit competence in the skills and habits of scientists. These processes include formulating scientific questions, designing experiments, collecting and analyzing data, developing explanations based on evidence, and communicating ideas with others (Rönnebeck et al., 2016).

Some assessments are 'practical' assessments in which students might engage in some element of a laboratory activity, such as focusing a microscope or taking a specific measurement, thus focusing on students' knowledge of science procedures (Li et al., 2006). Other efforts have involved what have been called 'performance assessments,' or assessments that engage students in using actual skills of science to respond to a particular question. For example, the 'Paper Towels' assessment (Shavelson et al., 1991) provides students with multiple brands of paper towels and different liquids and invites them to design an experiment to test which type of paper towel can absorb the most liquid.

Assessing content and inquiry together

Presenting these two elements separately, however, can misrepresent the way that science is actually done, as well as the ways reformers intend science to be learned in schools. Scientists draw on deep foundational knowledge of scientific theories and principles as they design and conduct experiments and communicate with others; similarly, scientists do not engage in the process of science without being informed by work that has been done previously. These two elements of science learning – assessing what students are able to do as they map the habits and processes of science, along with their understanding of science ideas – are at the core of international reform efforts in science education (e.g., OECD, 2015).

Thus, in order to be able to assess science education reforms as they are currently conceptualized, those tasked with the design of classroom assessments – curriculum designers and classroom teachers – must design what Baxter and Glaser (1998) called a process-content space, or a task which creates opportunities for students to simultaneously engage their participation in some kind of science practices or inquiry, as well as apply their science knowledge.

Current reform efforts in the US also seek to engage students in what have been called multidimensional learning experiences; that is, they engage students in science practices to learn disciplinary core ideas (NRC, 2012). In addition, the US approach also integrates crosscutting concepts, or those concepts such as scale, patterns, and systems that cut across science disciplines. Together, the disciplinary core ideas, science practices, and crosscutting concepts create three-dimensional learning goals in the *Next Generation Science Standards* (NGSS; NGSS Lead States, 2013) that fundamentally intertwine the process of science inquiry and science content.

Efforts to assess science content and inquiry in large-scale assessment

To explore approaches to assessing content and inquiry, it's useful to explore how international science assessments have approached them, as well as how those assessments have shifted over time. As a classroom teacher, I was squarely focused on my interactions with my own students and local standards and assessments. Indeed, scholars have called the isolation of classrooms an 'egg-crate,' in which teachers and students are, in a way, forming their own independent communities of learners, and teachers have significant amounts of autonomy in enacting learning experiences on a daily basis. However, what I encountered in my own classroom was influenced by larger movements in large-scale and international assessments, which not only reflect reform efforts locally, but also lead trends in what is valued in science education reform. Historically, large-scale, international assessments have also carried a great deal of weight in our conceptualization of how science should be learned in schools.

One of the first major efforts in this domain was the Third International Mathematics and Science Study (TIMSS), a standardized assessment first given to students in 41 countries in 1995. TIMSS included test items that primarily assessed science knowledge, as the item in Figure 10.1 illustrates.

This item in Figure 10.1 taps students' ability to recall different elements of the water cycle, and to place them in what is determined to be the 'correct' order

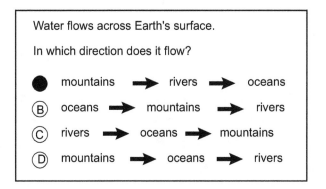

Figure 10.1 Sample TIMSS item

by the item designers. The TIMSS assessment has since been administered every four years to students in fourth and eighth grades in an increasingly large number of countries, and has also given rise to studies comparing curricula (Schmidt et al., 1997) and classroom teaching across participating countries (Stigler & Hiebert, 1999).

In 2000, the Organization for Economic Cooperation and Development (OECD) launched the Program for International Student Assessment (PISA). From its inception, PISA took a different framing than TIMSS, focusing on the integration of science knowledge and inquiry rather than assessing them separately. PISA has since been administered every three years to 15-year-olds in numerous countries, compelling significant shifts in educational policy within many participating nations (e.g., Hackling et al., 2017). The PISA approach is to assess science literacy through the lens of science competencies, and students are invited to apply their knowledge to in everyday scenarios (OECD, 2015). Figure 10.2 provides a sample of such an item, framed in the everyday context of fertilizers and plants.

Both assessments compelled countries to turn inward and critically self-examine how science was taught; many countries' attention was pulled to Finland and Taiwan, which topped the rankings. Germany, for example, was in the throes of the post-PISA 'shock' for many years, compelling new educational standards.

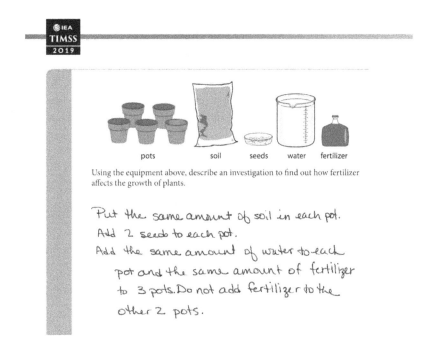

Figure 10.2 Sample PISA item assessing both science inquiry and content

In the US, which performed at the middling level in both assessments, the term 'a mile wide and an inch deep' was leveraged to criticize reform movements' emphasis on students learning so much content, but not in sufficient depth to be useful (Schmidt et al., 1997).

More recently, both the PISA and TIMSS assessments have shifted in their frameworks to more explicitly reflect the ways that scientists do their daily work, and how knowledge is generated in science disciplines. In this way, both assessments are seeking to move beyond the 'what' and toward the 'how' and 'why' of science. PISA's most recent science framework states that its goal is to assess scientific literacy defined as explaining phenomena scientifically, evaluating and designing scientific enquiry, and interpreting data and evidence scientifically. These three elements – called competencies – each require students to be able to recall and use the content knowledge they have learned in a particular everyday context or scenario.

A framework for tasks that assess content and inquiry together

As the preceding sections indicate, being clear about what we are intending to assess is a foundational first step in selecting, adapting, and designing assessments that tap into students' knowledge and abilities to engage in science practices. A useful framework for this process is the Assessment Triangle (NRC, 2001, 2014; Figure 10.3). The Assessment Triangle was originally developed for those working on large-scale assessments, but in my experience, it is also a helpful resource to get more clarity on what we are trying to assess as science teachers (e.g., Furtak et al., 2016).

The triangle helps to make explicit three key elements of assessment design – first, what is called the model of cognition, or a clear statement of what it is that is being assessed; second, the observation, or how information will be collected about a student's performance relative to that goal; and third, the interpretation, or the ways in which the student's observed performance will be compared to that goal.

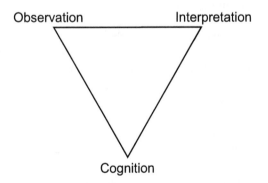

Figure 10.3 Assessment Triangle (NRC, 2001, 2014)

Let's talk about each of these in a little more detail, and how they apply to assessing students' engaging in multidimensional learning experiences (content + inquiry).

Cognition: what we are intending to assess

The triangle helps us think about how to assess content and inquiry together because it first asks for statements of what we are trying to assess. Rather than having an underlying approach of assessing just content or just inquiry, the goals then are written to combine both. Starting with this idea of a goal or construct that will assess both content and practice will help us be clear from the beginning what we are trying to support students in doing.

As examples of these kinds of goals or constructs, we can take the example of a three-dimensional performance expectation from the NGSS. For example, MS-ESS 1–1 describes that students will 'Develop and use a model of the Earth-Moon-Sun system to describe the cyclic patterns of lunar phases, eclipses of the sun and moon, and seasons.' In this example, the process or practice of science becomes the verb, with the student as the noun and the content as the direct object; that is, the model is developed to support students as they describe patterns in the Moon's phases and other observable phenomena. Thus, in order to demonstrate learning in alignment with this performance, students will have to not only understand how to develop and use a model in this context, but also to draw upon their relevant foundation knowledge of the components of the Earth-Moon-Sun system, and their relative positions and movements.

The OpenSciEd curriculum materials in the US provide another useful and easily accessible example I'll work with to illustrate how the triangle works (www.opensci ed.org). The middle school physical science unit on thermal energy is framed around the central question of 'How can containers keep stuff from warming up or cooling down?' The unit is designed to support students in learning about several three-dimensional performance expectations from the NGSS, which combine science practices with disciplinary core ideas and concepts that cut across disciplines.

To burrow down into a specific example of assessment, we can look at Lesson 6, which has as its learning goal:

> We complete an individual assessment that includes making predictions about whether a cup with a new lid design will keep a drink cooler than a cup with an old lid design, developing a plan for collecting data to see if the amount of liquid changed in either cup over time and developing a model to explain why one cup system would lose more mass than another. We figure out these things:
>
> • Liquids, gases, and solids are made of particles of matter.
> • Particles in a gas have a lot of space between them, but particles in liquids and solids do not.
> • Liquids and gases are made of particles that can move around freely, but solids are made of particles that cannot.

When we unpack this learning goal, we can see that its individual components combine several science practices (making predictions; developing a plan to collect data; developing a model), as well as conceptual knowledge (the structure of matter and differences between solids, liquids, and gases, and the particulate nature of matter). This example illustrates how learning goals can be written that intertwine both science practices and content, and thus set the foundation for designing a task that provides space for students to show what they know.

Observation: the design of the assessment task

The next vertex of the triangle invites us to consider how we intend to observe students as they engage in inquiry and demonstrate knowledge relevant to the model of cognition, or goal, that we started with. This means that whatever tasks we select, adapt, or design should provide space for students to illustrate not just what they know, but to provide evidence of how they are engaging in the practices or demonstrating competencies of science that are underlying the task.

So if our goal or model of cognition states that we seek to understand how students are able to develop and use a model to show how structural differences in the design of a lid for a beverage cup might make a difference in the amount of matter that is retained inside the cup, we can consider how the task itself would create space for students to both engage in these practices, as well as to draw upon their understandings of concepts.

When it comes to assessing both student engagement in science inquiry and their knowledge of content, international consensus has emerged that tasks should begin with everyday questions or scenarios (sometimes also called phenomena) that provide a target for students to engage their practices and content knowledge – in a sense, to engage in sensemaking (Furtak et al., 2021). A key way to engage students in each of these practices and content is to present them with an everyday scenario or phenomenon that they cannot immediately explain. As part of larger efforts to engage students in science learning that is applicable to their daily lives, this parallel effort invites students to also do this in the context of assessment.

It's important, however, to consider how the scenario being used for the assessment relates to instruction (e.g., Buell et al., 2019). For the PISA items described earlier, these scenarios are self-contained, where the context and necessary background information are built into the item. In the in the case of classroom assessments, teachers should consider the relationship between the scenario they are considering (e.g., is it similar to or the same as the scenario driving instruction, or is it different and thereby assessing students' ability to transfer their learning to a new context?).

To allow students to respond to these kinds of questions that create space to engage in science practices and to illustrate conceptual understanding, we need to think about designing tasks that go beyond asking one question, and rather contain multiple different components that provide space for students to engage in the complex cognitive activities we're asking them to do (NRC, 2014). This means that students might be provided with space to draw and label a model, describe a data collection plan, to write an explanation, and other elements. In addition, Kang et al. (2016) examined several different types of scaffolds that can help students to show what they know with these kinds of tasks, and found that providing checklists can help students know what to include in their model (e.g., what can be seen with their eyes, and also what cannot be seen), as well as drawings in combination with writing and sentence frames ('What I saw was _____,' 'Inside [the balloon] the particles were _____,' and 'I know this because _____,' p. 686).

These kinds of tasks have been identified as being at the highest level of cognitive demand in task frameworks that look both at the cognitive load that students need to engage in to respond to the task, as well as students' opportunity to engage in both science practices and content (Tekkumru-Kisa et al., 2015). Figure 10.4 illustrates the different kinds of tasks analyzed in the Task Analysis Guide in Science (TAGS) Framework (Tekkumru-Kisa et al., 2015), and differentiates between tasks that provide opportunities for students to engage in only science practices, to demonstrate knowledge of science content or, in the final column, to demonstrate integration of both content and practices (while noting that students are provided opportunities to 'do science' when there is less guidance and scaffolding provided in the structure of the task).

The OpenSciEd unit described earlier includes an assessment as part of Lesson 6 – about halfway through the unit – in which students are asked to explain how a company's decision to redesign its cold-beverage lids to no longer need straws might affect what happens to the liquid inside the cup. The goal states

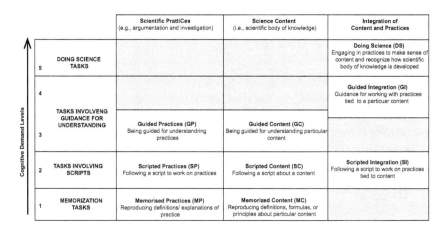

Figure 10.4 Task Analysis Guide in Science (Tekkumru-Kisa et al., 2015)

that students will have the opportunity to explain an approach to collecting data relevant to this question, as well as to draw a model that illustrates their understandings of phases of matter and molecular motion (see Figure 10.5).

☑ Explaining the Effect of Different Lid Designs

A beverage company wants to stop using plastic straws, so the company designs a new lid. The new lid has a much bigger hole than the old lid.

OLD LID   NEW LID

Alex, a scientist at the company, is wondering how this might affect what happens with the liquid inside the cup.

1) First, Alex does an experiment in which she fills 2 cups with cold liquid at the same temperature: 1 cup with the old lid and 1 cup with the new lid. She finds that the liquid in one cup warmed up more than the liquid in the other cup after 10 minutes. Which cup do you think warmed up more? Why?

3) At the end of 30 days Alex finds that one cup lost more liquid than the other. Add particle diagrams for the zoomed-in views at lid locations A and B to help explain how that could happen.

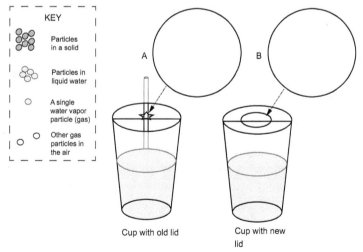

4) How do your diagrams help explain how one cup could lose more liquid than the other?

Figure 10.5 Sample questions from OpenSciEd task Lesson 6 Assessment

It's important to note that students need support in learning to respond to tasks such as these, highlighting the importance of using scaffolds and other approaches to help them learn what goes into these kinds of responses over time. In order to support students in showing us what they know and engaging both their knowledge of science concepts as well as their understandings of science inquiry, we often need to provide them with scaffolds in order to do so. As Tekkumru-Kisa et al. (2015) noted, this scaffolding may be necessary in order to support students in showing both their content and inquiry. Similarly, reaching what some have articulated as top-level competence in some science inquiry domains, such as spontaneously drawing models when presented with a new phenomenon, necessitates students learning over time what goes into this practice.

Interpretation

The final element of the Assessment Triangle encourages us to compare the evidence collected about what students know and are able to do in the *observation* back to the original goal or model of *cognition*. Here it is essential to consider guiding frameworks for assessing student ideas that keep together the elements that were originally articulated in the goal/construct, and to not focus assessment and feedback on what might be easier to score (e.g., the content). Rather, looking at developing frameworks or rubrics or progressions that show how student competence develops over time can help to guide us in, for example, evaluating student responses relative to the position of the assessment in the unit (e.g., students may not be expected to have developed full competency in meeting a unit learning goal midway through the unit; rather, this point may be better for determining what students have learned so far, and what they need to continue to work on going forward).

Lesson 6: Rubric

Rubric for Model in Lesson 6 Assessment

Components	Category			Feedback
The model clearly represents or describes the system components and includes:	Missing	Developing	Mastered	
• water particles in the air on the inside and outside of the lid of the cup				
• system boundaries in the form of particles that make up the solid material of the lid				
• water particles moving through the air				
Interactions between Components	**Category**			**Feedback**
The model clearly represents or describes the following:	Missing	Developing	Mastered	
• solid particles with spacing between them (that is less than the width of the diameter of a water particle) to represent the edge of the lid near the opening *(pattern/structure)*				
• a gap in the solid particles to represent the opening or the start of the opening *(pattern/structure)*				

Figure 10.6 Rubric for OpenSciEd Assessment at Lesson 6

Rubrics are a useful example to guide teachers' interpretation of student responses to assessments that combine both practices and concepts that can unpack and provide space for feedback to students on different elements of the underlying learning goal. For example, the Lesson 6 Assessment from OpenSciEd includes a rubric that teachers may use to examine how models represent components of the water-cup system, as well as interactions between components (Figure 10.6). The rubric provides space for teachers to identify what students have included in their work, as well as what still needs to be included, and space to provide feedback. This kind of written feedback has been illustrated as an essential element to supporting students' learning, particularly in the context of formative assessment for inquiry-based learning (e.g., Holmeier et al., 2018).

Assessment planner: criteria to consider when designing an assessment for content and process

As the preceding steps illustrate, there are a number of key considerations to take into account when designing, selecting, and/or adapting tasks that effectively provide space for students to show their knowledge of both science inquiry and concepts. These are summarized in Table 10.1 for quick reference.

Table 10.1 Planning guide for assessments of content and science inquiry

Assessment Triangle	Planning step	Questions to consider
Model of cognition	Learning goal(s)	What do I want to find out about my teaching of the lesson and/or student learning? What elements of science inquiry/practice will students engage in? What big ideas/concepts will they need to know?
Observation	Scenario/phenomenon	What accessible/observable event will be used to frame the assessment task?
	Task elements	What kinds of question formats are necessary to provide space for students to demonstrate their knowledge of the learning goal? What kinds of supports and scaffolds are provided to help students share what they know?
Interpretation	Rubrics/scoring guides	What interpretive frameworks are available, or can I plan in advance, to help me make interpret student responses relevant to the learning goal?
	Feedback	What is the plan to provide substantive feedback to students to support their ongoing learning?

Teachers may use this overall framework to work together to identify, select, adapt, or – in some cases – design assessment tasks that will meet learning goals that integrate both science concepts and inquiry. Teachers working together to engage in this kind of collaborative work over time can be supported in making sense of the demands framed in new science teaching reforms (e.g., Heredia, 2020), as well as to design better assessment tasks over time (Furtak et al., 2016).

Conclusion

This chapter has described the contrast between students' conceptual knowledge and their engagement in the skills and thinking processes of scientists – also called science inquiry – that has been in existence for a long time. Just as these debates have arisen in the context of curriculum design and discussions of student learning for years (Rudolph, 2002), they have also manifested in international large-scale assessments. The structure of many school systems means that what is assessed is what is taught. As such, it is fundamental that as visions for science learning shift to integrate both content and science inquiry, that assessments mirror these visions for classroom science learning.

Questions for further debate

1   What should the balance be between assessing knowledge, understanding, and skills in science education?
2   How can we improve students' understanding of the purposes and practices of assessment?

Suggested further reading

Molnar, G. (2021). Challenges and developments in technology-based assessment: Possibilities in science education. *Europhysics News, 52*(2), 16–19.

Rustaman, N. Y. (2017). Assessment in science education. *Journal of Physics: Conference Series, 895* (International Conference on Mathematics and Science Education (ICMScE) 24 May 2017, Bandung, Indonesia).

References

American Association for the Advancement of Science. (1990). *Science for all Americans.* Oxford University Press.

Baxter, G. P., & Glaser, R. (1998). Investigating the cognitive complexity of science assessments. *Educational Measurement: Issues and Practice*, (1), 37–45.

Briggs, D. C., Alonzo, A. C., Schwab, C., & Wilson, M. (2006). Diagnostic assessment with ordered multiple-choice items. *Educational Assessment, 11*(1), 33–63.

Buell, J., Furtak, E. M., Deverel-Rico, C., & Henson, K. (2019, April). *Toward a framework for selecting phenomena at the intersection of curriculum and assessment.* Paper presented at the annual meeting of the National Association of Research in Science Teaching, Baltimore, MD., April 2, 2019.

Chen, Z., & Klahr, D. (1999). All other things being equal: Acquisition and transfer of the control variables strategy. *Child Development, 70*(5), 1098–1120.

Ford, M. J. (2015). Scientific practices educational implications of choosing "practice" to describe science in the next generation science standards. *Science Education, 99*(6), 1041–1048. https://doi.org/10.1002/sce.21188

Ford, M. J., & Forman, E. A. (2006). Redefining disciplinary learning in classroom contexts. *Review of Research in Education, 30*(1), 1–32.

Furtak, E. M., Badrinarayan, A., Penuel, W. R., Duwe, S., & Patrick-Stuart, R. (2021). Assessment of crosscutting concepts: Creating opportunities for sensemaking. In J. Nordine & O. Lee (Eds.), *Crosscutting concepts: Strengthening science teaching* (pp. 333–356). NSTA Press.

Furtak, E. M., Glasser, H., & Wolfe, Z. (2016). *The feedback loop: Using formative assessment data to inform science teaching and learning*. NSTA Press.

Furtak, E. M., Kiemer, K., Circi, R. K., Swanson, R., de León, V., Morrison, D., & Heredia, S. C. (2016). Teachers' formative assessment abilities and their relationship to student learning: findings from a four-year intervention study. *Instructional Science, 44*(3), 267–291.

Hackling, M. W., Ramseger, J., & Chen, H. S. (2017). *Quality teaching in primary science education*. Springer.

Heredia, S. C. (2020). Exploring the role of coherence in science teachers' sensemaking of science-specific formative assessment in professional development. *Science Education, 104*(3), 581–604.

Holmeier, M., Grob, R., Nielsen, J. A., Rönnebeck, S., & Ropohl, M. (2018). Written teacher feedback: Aspects of quality, benefits and challenges. In *Transforming assessment: Contributions from science education research* (vol. 4, pp. 175–208). Springer. https://doi.org/10.1007/978-3-319-63248-3

Kang, H., Windschitl, M., Stroupe, D., & Thompson, J. (2016). Designing, launching, and implementing high quality learning opportunities for students that advance scientific thinking. *Journal of Research in Science Teaching, 53*(9), 1316–1340.

Klahr, D., & Nigam, M. (2004). The equivalence of learning paths in early science instruction. *Psychological Science, 15*(10), 661–667.

Latour, B., & Woolgar, S. (1979). *Laboratory life: The construction of scientific facts*. Princeton University Press.

Li, M., Ruiz-Primo, M. A., & Shavelson, R. J. (2006). Towards a science achievement framework: The case of TIMSS-R study. In T. Plomp & S. Howie (Eds.), *Contexts of learning mathematics and science: Lessons learned from TIMSS* (pp. 291–312). Routledge.

National Research Council. (2001). *Classroom assessment and the national science education standards*. National Academy Press.

National Research Council. (2012). *A framework for K-12 science education: Practices, crosscutting concepts, and core ideas*. The National Academies Press.

National Research Council. (2014). *Developing assessments for the next generation science standards*. National Academies Press.

NGSS Lead States. (2013). *Next generation science standards: For states, by states*. The National Academies Press.

OECD. (2017). *PISA 2015 assessment and analytical framework: Science, reading, mathematic, financial literacy and collaborative problem solving* (revised ed.). OECD Publishing.

Otero, V., & Nathan, M. J. (2008). Preservice elementary teachers' views of their students' prior knowledge of science. *Journal of Research in Science Teaching, 45*(4), 497–523.

Rönnebeck, S., Bernholt, S., & Ropohl, M. (2016). Searching for a common ground – A literature review of empirical research on scientific inquiry activities. *Studies in Science Education, 52*(2), 161–198.

Rönnebeck, S., Nielsen, J. A., Olley, C., Ropohl, M., & Stables, K. (2018). The teaching and assessment of inquiry competences. In *Transforming assessment* (pp. 27–52). Springer.

Rudolph, J. L. (2002). *Scientists in the classroom: The cold reconstruction of American science education*. Palgrave.

Sadler, P. M. (1998). Psychometric models of student conceptions in science; reconciling qualitative studies and distractor-driven assessment instruments. *Journal of Research in Science Teaching, 35*(3), 265–296.

Schmidt, W. H., McKnight, C. C., & Raizen, S. A. (1997). *A splintered vision: An investigation of U.S. science and mathematics education*. Kluwer Academic Publishers.

Schwab, J. J. (1962). *The teaching of science as enquiry*. Harvard University Press.

Shavelson, R. J., Baxter, G. P., & Pine, J. (1991). Performance assessment in science. *Applied Measurement in Education, 4*(4), 347–362.

Shepard, L. A. (2000). The role of assessment in a learning culture. *Educational Researcher, 29*(7), 4–14.

Stigler, J. W., & Hiebert, J. (1999). *The teaching gap: Best ideas from the world's teachers for improving education in the classroom*. The Free Press.

Tekkumru-Kisa, M., Stein, M. K., & Schunn, C. (2015). A framework for analyzing cognitive demand and content-practices integration: Task analysis guide in science. *Journal of Research in Science Teaching, 52*(5), 659–685.

Chapter 11

Science pedagogies from an international and comparative perspective

Ann Childs

Introduction

This chapter uses two major international student assessments, the Trends in International Mathematics and Science Study (TIMSS) and the Programme for International Student Assessment (PISA), to explore key issues about science pedagogies internationally from a comparative perspective. These assessments have proved to be very influential, and the results from TIMSS and PISA have had considerable impact on whole countries' educational policies, often described as 'PISA shock'. For example, Germany's PISA shock occurred after the results of PISA 2000 were published:

> The publication of the first round of the Programme for International Student Assessment (PISA) results in December 2001 had a "Tsunami-like impact" (Gruber, 2006, p. 195) in Germany, dominating news headlines for weeks and transforming educational policy-making discourse with effects that are still discernible today. The 'PISA-shock' had been preceded by the considerably less violent 'TIMSS-shock' four years earlier, which made the public aware of international large-scale assessments.
>
> (Waldow, 2009, p. 476)

The effect in Germany, as Sjøberg and Jenkins (2020) argue, was widely felt at the political level in the German election of 2007, and also led to a number of major initiatives to improve school science and mathematics. Similarly, Australia also experienced its own PISA shock in 2010, with students in China (Shanghai) outperforming Australian students by a wide margin (Baroutsis & Lingard, 2017). In August 2012, Julia Gillard, then-Prime Minister of Australia, announced that Australia would 'strive to be ranked in the "top five" in international education assessments by 2025' (Gorur & Wu, 2015, p. 648), and this ambition became part of the Australian Education Act of 2013. Finally, Norway underwent its own PISA shock in 2000, which lead to the introduction of a national quality assessment system, national tests, changes to the mathematics curriculum and to the education of mathematics teachers (Nortvedt, 2018).

DOI: 10.4324/9781003137894-13

Therefore, these tests matter and so debates rage about what, if anything, lower-performing countries can learn about effective science teaching and learning and the education of teachers in the countries that perform highly in PISA and TIMSS, particularly if they are culturally very different from their own. This chapter explores the following questions:

- What do the findings from TIMSS and PISA say about science teaching and learning internationally?
- What do the findings from TIMSS and PISA suggest about the influence of culture and context on science teaching and learning?
- What can countries internationally learn from TIMMS and PISA to develop science teaching and learning in their own countries?

The chapter begins by briefly describing how the two major assessments, TIMSS and PISA, work. It then goes on to address the three questions raised before offering some concluding thoughts about science teaching and learning across the world and debates about the ways these assessments can inform the development of science teaching and science teacher education internationally.

TIMSS and PISA: Trends in International Mathematics and Science Study (TIMSS)

The International Association for the Evaluation of Educational Achievement (IEA), based in the Lynch School of Education, Boston College, USA, administers TIMSS. TIMSS is a series of international assessments of the mathematics and science knowledge of students from diverse countries with different levels of economic development and population size. TIMSS was first carried out in 1995, although it had some precursors. For example, in 1964, the First International Mathematics Study (FIMS) was carried out by the IEA with 11 countries participating with students aged 13. In 1970, the First International Science Study (FISS) for students aged 10, 14 and the final year of secondary education, which then took place again between 1980 and 1984. In 1995, TIMSS was conducted at five grade levels (third, fourth, seventh and eighth grades, and the final year of secondary school) in mathematics and science in 40 countries. In addition, information was collected from students, teachers and head teachers about the teaching and learning of mathematics and science, and information about the curricula of the participating countries was also collected.

TIMSS 1999 tested eighth-grade students only in mathematics and science, and additional information on teaching and learning and curricula were also collected. In addition, TIMSS 1999 included a video study of eighth-grade lessons in mathematics and science in the USA, Australia, the Czech Republic, Hong Kong SAR, Japan, the Netherlands and Switzerland. TIMSS 2003 again tested mathematics and science for both fourth-grade and eighth-grade students. There were, however, some changes in the nature of the questions, which placed more emphasis on enquiry and problem-solving. Additional background information was again collected on the quantity, quality and content of instruction.

In 2007, 2011, 2015 and 2019, fourth- and eighth-grade students were now assessed. In 2019, the seventh cycle of TIMSS, the transition moved to a digital assessment format, where half the countries participating did so digitally, and the other half continued with the paper-based tests. The next round of testing will take place in 2023. TIMSS also has an advanced assessment for students in their final year of secondary education. TIMSS Advanced was first conducted in 1995, then again in 2008 and 2015. It assesses final-year secondary students' achievement in advanced mathematics and physics in order to give information about a student's readiness for transition to university. Additional information about other contextual issues, such as teacher training, curriculum and technology use, was also collected.

Programme for International Student Assessment (PISA)

Like TIMSS, the Programme for International Student Assessment (PISA) is an international test conducted by the Organisation for Economic Development (OECD) in both member and non-member countries of the OECD. PISA first began in 2000 and tested 15-year-old school students' performance in mathematics, science and reading and is repeated every three years. While PISA always tests reading, mathematics and science for every year it has taken place, one of these areas is chosen as a focus. Table 11.1 shows the focus for each of the different years in which PISA was conducted, and it also gives information on additional data collected.

Table 11.1 Summary of the PISA foci and additional data collected 2000–2018

Year	Focus	Additional data
2000	Reading	Schools provided background information.
2003	Mathematics	Questionnaires to students and principals on social, cultural, economic and educational factors.
2006	Science	Questionnaires to students and principals about their schools. In 16 countries, parents completed a questionnaire about their children's education and their views on science related issues and careers.
2009	Reading	• Questionnaires to students about themselves and their homes and questionnaires to principals about their schools. • In some countries, optional short questionnaires to (i) parents providing information on students' engagement with reading at home and (ii) students to provide information on access to and use of computers (also educational history and aspirations).

(Continued)

Table 11.1 (Continued)

Year	Focus	Additional data
2012	Mathematics	Questionnaires to students and principals about students' backgrounds, schools and learning experiences. In addition, also about the school system and the wider learning environment.
2015	Science Optional assessment on students' financial literacy	• Questionnaires to students giving information about themselves, their homes, and their school and learning experiences. • Questionnaires to principals about the school system and the learning environment. • Some countries gave optional questionnaires to parents on their perceptions of and involvement in their child's school, their support for learning in the home and their child's career expectations, particularly in science. • Two optional questionnaires for students: (i) information about familiarity with and use of information and communication technologies and (ii) information about their education, including any disruption to their schooling and thoughts about career preparation.
2018	Reading – delivered by computer in most of the 79 countries which took part Optional assessment on students' financial literacy	Questionnaires to students on attitudes and wellbeing.

PISA and TIMMS – similarities and differences

As these sections show, there are some similarities between TIMSS and PISA, but there are also some significant differences. Sahlberg (2011) explores these differences when he says:

Firstly, TIMSS measures how well students have learned different areas of the school curriculum, in other words, knowledge and skills included in mathematics and science teaching. PISA, in addition, focuses on how well students at the beginning of upper secondary school are able to use the knowledge and skills they have learned in new situations. Secondly, the IEA studies include a varying number of countries in four-year cycles, whereas PISA is primarily designed for developed OECD member countries, all of which have participated in every three-year cycle since 2000. Finally, TIMSS and PIRLS examine pupils who are in the 4th and 8th grades regardless of their ages, whereas students taking PISA tests are all 15 years old at the time of the tests. This means that IEA assessments are also able to follow up the age cohorts of 4th graders from one cycle to the next, whereas PISA does not.

(pp. 120–121)

These differences merit further examination in one key area. Sahlberg identified that TIMSS' focus for testing is on the school curriculum, but PISA's is on students' literacy, mathematical literacy and scientific literacy as per Lingard and Grek (2007).

Its innovative dimension lies in the fact that rather than examining students' mastery of the school curricula, the focus is on an assessment of young people's ability to practically apply their skills in everyday situations. PISA's definition of scientific literacy is as follows:

> PISA's assessment of students' scientific knowledge and skills is rooted in the concept of *scientific literacy*, defined as the extent to which an individual:
>
> • Possesses scientific knowledge and uses that knowledge to identify questions, acquire new knowledge, explain scientific phenomena and draw evidence-based conclusions about science-related issues.
> • Understands the characteristic features of science as a form of human knowledge and enquiry.
> • Shows awareness of how science and technology shape our material, intellectual and cultural environments.
> • Engages in science-related issues and with the ideas of science as a reflective citizen.
>
> (OECD, 2007, p. 12)

Some academics have raised concerns about the challenges in designing questions to test students' scientific literacy and have questioned "the extent to which PISA has been successfully able to devise questions that apply to the everyday cultural contexts of 15-year-olds from Hungary to China and from Canada to Uruguay has been disputed" (Lingard & Grek, 2007, pp. 15–16). This view is also supported by Sjøberg and Jenkins (2020), but they go further and question whether PISA data can take account of "the many different beliefs, assumptions, pedagogical practices, and cultural, social, economic and political contexts within which schooling takes place" (p. 11). Sjøberg and Jenkins (2020) and El Masri et al. (2016) have also demonstrated the considerable challenges in translation of these tests into the diverse languages of the participating countries.

Finally, as can be seen in the overviews of TIMSS and PISA, there are incredibly rich data sets which report on students' achievement in science, as well as collecting a wealth of other data from students, teachers, principals and parents. Hopfenbeck et al. (2018) conducted a systematic review of the number of academic journal articles published using PISA data between 2000–15 and identified articles where authors had drawn on PISA data and categorised these into three key groups: secondary analysis, policy analysis and critiques of PISA. They found that:

> the majority of articles, those in secondary data analysis, use PISA data as a foundation from which to build additional levels of newly constructed knowledge to bolster even further the already vast amount of information generated by PISA alone.
>
> (p. 347)

Similar secondary analyses have also been done with TIMSS data (see for example Drent et al., 2013), and it is these secondary analyses and other literature to which the following sections turn to answer the key questions in the introduction.

What do the findings from TIMSS and PISA say about science teaching and learning internationally?

A key finding found consistently in many studies and reports is that high scores in science assessments in PISA and TIMSS correlate negatively with inquiry-based teaching (see Areepattamanni, 2012; Forbes et al., 2020; Lau & Lam, 2017; McConney et al., 2014; Oliver et al., 2019; Sjøberg, 2015, 2018, 2019; Sjøberg & Jenkins, 2020). For example, Oliver et al. (2019) did a secondary analysis of PISA 2015 data for Australia, Canada, Ireland, New Zealand, the UK and the USA. Their secondary analysis investigated the efficacy of inquiry-based teaching in contrast with two other approaches to teaching science: teacher-directed teaching and adaptive teaching (where teachers adapt their instruction to meet the needs of learners). They found that "students who reported experiencing high frequencies of inquiry strategies in their classrooms consistently evidenced lower levels of scientific literacy across the six countries" (p. 1).

This finding is also consistent with their secondary analysis of the PISA 2006 data (see McConney et al., 2014). Sjøberg (2019) reports similar findings: "For the variation within the same country, the PISA finding is that in no education system do students who reported that they are frequently exposed to enquiry-based instruction . . . score higher in science" (p. 29). Furthermore, Sjøberg (2019) also identifies that in higher-performing countries like Japan, Korea, Taiwan, China (Shanghai) and Finland, which he calls the 'PISA winners', students also reported very little use of inquiry-based teaching. As many writers argue, these findings challenge the current received wisdom from many influential science educationalists and scientists who advocate inquiry-based teaching in science (Sjøberg, 2015). For example, as influential science educator Wynne Harlen (2013) says:

> [T]he value of IBSE is not a matter that can be decided by empirical evidence, but it is a value judgement that the competences, understanding, interest and attitudes that are its aims are worthwhile and indeed are necessary in a modern education.
>
> (p. 4)

However, although this is a consistent headline finding, there is also more nuance and complexity identified in these secondary analyses. First, in a secondary analysis of TIMSS 2015 data, Teig et al. (2018) found that inquiry-based teaching did not necessarily result in lower achievement; rather, it was the amount of inquiry-based teaching that was at issue:

> The increasing use of inquiry-based teaching was correlated with higher achievement in science until it reached an optimum value, then this association decreased as the use of the strategy increased. Furthermore, when there was a very high frequency of inquiry activities, the relationship between inquiry and achievement became negative.
>
> (p. 27)

Second, Jerrim et al. (2019), in a secondary analysis of 2015 PISA data of students from England, found again that the amount of inquiry-based teaching students were exposed to had an effect on attainment, and they also found that the level of guidance in the inquiry was significant:

> [I]nquiry-based teaching has a very weak relationship with attainment in science – and that any positive effects are confined to moderate levels of inquiry combined with high levels of guidance. High levels of inquiry or unguided inquiry have no relationship with attainment at all. These results are consistent with existing literature, which tends to find that inquiry is less effective than more direct forms of instruction (Alfieri et al., 2011; Kirschner et al., 2006; Stockard et al., 2018) except for in cases where the inquiry is highly guided (Hmelo-Silver, Duncan, & Chinn, 2007; Lazonder & Harmsen, 2016).
>
> (p. 42)

Third, Oliver et al. (2019) did a very fine-grained secondary analysis of the data, where they looked at the more individual elements that make up inquiry-based teaching. PISA 2015 also included surveys of students' perceptions of the teaching and learning strategies they experienced, and for inquiry-based teaching these included:

> (1) students are given opportunities to explain their ideas; (2) students spend time in the laboratory doing practical experiments; (3) students are required

to argue about science questions; (4) students are asked to draw conclusions from an experiment they have conducted; (5) the teacher explains how a science idea can be applied to different phenomena; (6) students are allowed to design their own experiments; (7) there is a class debate about investigations; (8) the teacher clearly explains the relevance of science concepts; and (9) students are asked to do an investigation to test ideas.

(OECD, 2016, p. 8)

Taking these more individual elements of inquiry-based teaching, they found three types of association between students' scientific literacy and how frequently they reported experiencing the strategies listed here. First, there were some inquiry-based teaching strategies, which seemed to have no association with students' scientific literacy; for example, where students were given opportunities to explain their ideas. Second, there were inquiry-based teaching strategies, which seemed to have a negative relationship with students' scientific literacy. So, for the two instructional strategies, 'experience class debates in science' and 'arguing about science questions', higher frequencies of these strategies seemed to be associated with lower levels of scientific literacy. Oliver et al. (2019) also found a third type of association:

> This type can be characterised as non-linear (curvilinear). Two examples are the items that ask students about the frequency with which they experience spending time in the laboratory doing practical experiments (in our view, a sine qua non of inquiry-based science education) and the item asking how often students experience drawing conclusions from an experiment they have conducted, a widely recognised critical aspect of inquiry-oriented learning and teaching in science (Furtak et al., 2012; McConney et al., 2014; Minner et al., 2010). In the first case [. . .] for each country, the highest level of students' scientific literacy is associated with spending time doing practical experiments in some lessons, rather than in most or all lessons. Similarly, in drawing conclusions from an experiment they have conducted (Fig. 4, middle right), in all six countries, higher levels of scientific literacy are associated with students engaged this activity in some lessons (Canada, New Zealand, UK) or in most lessons (Australia, Ireland, USA) rather than in all lessons or never. This more nuanced, non-linear patterning was remarkably consistent across the six countries included in this analysis.

(p. S610)

As well as giving a more fine-grained and nuanced analysis into the different elements of inquiry-based teaching and their association with developing scientific literacy, again these findings suggest that the frequency and amount of time spent on certain elements of inquiry-based teaching is important to consider.

Lau and Lam (2017) in their secondary analysis of PISA 2015 data, also found similar issues in the literature in the area of practical work and science investigations which did not seem to be beneficial to PISA performances and so, like Oliver et al. (2019), this poses questions about the value as well the amount of practical work to which students should be exposed. However, for Lam and Lau, this did not mean that investigations or practical work should be abandoned; rather, they perceived that these findings should alert us to the need to engage in how to make practical work more effective. Indeed, concerns about the amount and value of practical work are not limited to these secondary analyses of PISA and TIMSS data; other research has raised similar concerns too (see for example Abrahams, 2009, 2011; Abrahams & Millar, 2008).

In summary, these more fine-grained findings from PISA resonate with Teig et al.'s (2018) findings from TIMSS and show that it is not that inquiry-based teaching per se which is at issue; rather, it might be more complex and involve factors like the amount of inquiry-based teaching, the amount of teacher guidance and that different elements of inquiry-based teaching give different scientific literacy outcomes. These findings seem to suggest that further research needs to be conducted, and I will come back to this point later. Furthermore, the findings also suggest that teacher-directed and adaptive approaches are associated with high-performing countries, and as Sjøberg (2019) reported, the top-performing countries' students reported very little inquiry-based teaching. The picture is complex.

Finally, many of these secondary analyses look at the association between inquiry-based instruction and student interest and engagement with science. McConney et al. (2014), in their analysis of PISA 2006 data, found that high levels of inquiry-oriented learning activities were associated with "above-average levels of interest in learning science and above average engagement with science" (p. 963). Sjøberg (2019) and Sjøberg and Jenkins (2020) comment on this issue and similarly note that students in the high-performing countries also seem to develop the most negative attitudes to science. As Sjøberg (2019) says:

> The possibly most surprising finding is that many countries with the highest mean PISA science scores were at the bottom of the list of students' interest in science. Finland and Japan are prime examples: at the top on PISA science score, and at the very bottom on constructs like "interest in science", "future-oriented motivation to learn science" as well as on "future science job", i.e. inclination to see themselves as scientists in future studies and careers.
> (Sjøberg, 2019, p. 27)

Sjøberg and Jenkins (2020) go on to argue that this finding suggests a "need for a deeper understanding of the relationship between PISA science scores and measures of student attitudes and interest" (p. 4). Zhao (2017), in looking at results from both PISA and TIMSS, also notices this tension between students' attainment, their interest and engagement in science in high-performing countries. He makes the very pertinent point that "if indeed the policies and practices that raise test scores also hurt confidence and attitude, we must carefully weigh the risks against their benefits. Do we care more about test scores or confidence or attitude?" (p. 11). However, as science teachers, do we really have to sacrifice one for the other? I will come back to this point later.

What do the findings from TIMSS and PISA suggest about the influence of culture and context on science teaching and learning?

The answer to this question is also complex and, in this section, I will draw on secondary analyses from the PISA 2015 study (see for example Lau & Lam, 2017), as well as the findings from the 1999 TIMSS video study, to address this question (Roth et al., 2006).

First to Lau and Lam (2017). This study is again a secondary analysis of PISA 2015 data which analysed the science performance and instructional strategies used by the 10 top-performing regions in PISA 2015. The 10 regions were Singapore, Japan, Estonia, Taipei, Finland, Macao, Canada, Hong Kong, China and Korea, and one of their key research questions was, "Are the science teaching practices, science performances, and their relationships associated with the cultures of these regions?" (p. 2129).

In their research they classified the different cultural groups as follows: "East Asian (Japan, Korea, Singapore), Chinese (B-S-J-G, Hong Kong, Chinese Taipei and Macao), Western (Canada) and Nordic (Finland and Estonia)" (p. 2129), and within the Chinese cultural group, they also identified those influenced by Confucian Heritage Culture and suggested that Japan and Korea fall under this umbrella. They also investigated the similarity between teaching practices and performance and how science teaching practices related to science performance. Their findings across all these areas are rich, complex, detailed and interesting to read. However, in relation to the specific question asked, they found that:

> No cultural clusters could be identified. For example, Canada, Singapore and Finland have the most teacher-directed instruction in science class, while feedback given to students is more common in Singapore, China and Chinese Taipei. It is more common in the Western classroom that students are allowed to express their ideas, but China is similar to the Western regions in that aspect. In general, Canada and Singapore tend to be more student- and enquiry-oriented in science teaching, while Japan and Korea are more traditional and didactic.
>
> (p. 2144)

In relation to the other areas of focus, they found again that high performance in science was associated with a high frequency of teacher-directed instruction, except in Japan and Korea, but they also found that there was a "fair amount" of practical work done with some investigations (p. 2144). In terms of performance and teaching strategy, they concluded that "there is no single success formula for science education as shown by the findings of the study" (p. 2145).

Second, the TIMSS 1999 video study (Roth et al., 2006), conducted in the USA, Australia, the Czech Republic, Hong Kong SAR, Japan, the Netherlands and Switzerland, also attempted to investigate whether "the science education community of researchers and educators could learn anything about teaching strategies as they are actually practiced across cultures" (Roth et al., 2006, p. 5). The findings resonate to a certain extent with Lau and Lam when this study found that "no single approach was shared by the four higher-achieving countries" (p. 22). Therefore, like Lau and Lam, there was no single success formula to be able to draw upon. However, they did find some commonalties in the higher-achieving countries, which were "all characterized by a core instructional approach that included a consistent instructional and content organization strategy that held students to some type of high content standards" (p. 22). But they also found that each country had distinct approaches to science teaching which provided students with different opportunities to learn science "and different visions of what it means to understand science" (p. 22). For example, the analysis categorised Japan's approach as 'Making Connections between Ideas and Evidence'. The Netherlands was characterised as 'Learning Science Independently', and the Czech Republic as 'Talking about Science Content'. This challenges Lau and Lam's findings to a certain extent in that distinct approaches in different countries and cultures could be identified, but does it provide a blueprint for other countries/science educationalists to learn from? Lau and Lam (2017) caution us to think about how culture and social values interact with many other factors to influence a country's pedagogy and the achievement of its students:

> As for the roles of cultural and social values in science teaching, caution is needed to attribute the top performance of these regions to particular teaching practices, which are further attributed to particular cultural and social values. There is a host of factors at play other than teaching practices in affecting the educational outcomes of a region, such as demographics, school system, education policy and teacher training.
>
> (p. 2145)

So, for example, Finland's success in TIMSS and PISA has consistently been attributed to, amongst many factors, the high social inclusion of all students in the education system, the high qualifications of its teachers, the high status and trust in which teachers are held by the society and the relative autonomy of teachers to exercise their own professional judgement with light-touch accountability measures on teachers (see for example Sahlberg, 2011, Ustun & Eryilmaz (2018). However, in looking at these secondary analyses of teaching in Finland, it is also found that the more conservative and authoritarian culture in Finland means that teachers tend to adopt more conservative teacher-directed approaches. Simola (2005) has attributed this to the background of Finland being rooted in past history and influences from the Russian Empire, which promoted an authoritarian, compliant and collectivist approach to Finnish culture that still influences teaching and learning in Finland. As Simola says:

> To put it simply, it is still possible to teach in the traditional way in Finland because teachers believe in their traditional role and pupils accept their traditional position. Teachers' beliefs are supported by supported by social trust and their professional academic status, while pupils' approval is supported by the authoritarian culture and mentality of obedience.
>
> (p. 465–6)

Therefore, to understand Finland we have to, as Lam and Lau argue, understand the myriad complex and interacting factors that influence teaching and learning, which include a country's culture that is often rooted in its social and historical past. Given this complexity and the issues and challenges of understanding each jurisdiction and each country's unique circumstances, the next section examines the extent to which it is possible, or even desirable, to learn about effective pedagogical approaches in science from high-performing countries in TIMSS and PISA.

What can countries internationally learn from TIMSS and PISA to develop science teaching and learning in their own countries?

There are considerable debates about the ways and extent data from TIMSS and PISA can inform educational policy and development in lower-performing countries. Feniger and Lefstein (2014), focusing their critique on PISA, call this assumption 'PISA reasoning':

> A central assumption – perhaps the central assumption – underlying PISA comparative work is what we call the 'Policy Structures Assumption', i.e. global variation in students' educational performance is primarily attributable to national educational policies.
>
> (p. 846)

Indeed, in the introduction to this chapter, it can be seen that many counties have overhauled their whole education systems as a result of the so-called PISA shock. Feniger and Lefstein offer a different assumption, the Cultural-Historical Assumption, which "attributes differences between average student scores to historical trajectories of social, economic and educational development and to cultural values, beliefs and practices" (p. 847). In Gorur and Wu's (2015) analysis of Australia's performance in PISA, they also argue that social and cultural histories are important in understanding a country's performance and made a number of interesting points specific to the context of Australia, but which may apply more widely. First, they identified a huge appetite in Australia for borrowing from the policies and practices of high-performing countries in PISA, and they caution that simply concluding that these strategies are effective could be dangerous when the same practices could also be found in lower-performing countries. Feniger and Lefstein (2014) made a similar point when they said:

> X may well be a common feature of high-performing education systems a, b, c, d and e, but that doesn't demonstrate a cause-effect relationship between feature and performance. And if x is also a common feature of low-performing systems g, h, i, j and k, then the claimed correlation is clearly inadmissible.
>
> (p. 8)

Second, Gorur and Wu (2015) expressed concern that the high performance of students in the East Asian systems, to which Australia was being compared, may be due to parents investing significantly in private tutoring or in after-school private cram-schools, and concerns that for Australia, this practice has been shown to have detrimental effects on students' wellbeing and mental health. Feniger and Lefstein (2014) echo this point and attribute the success of East Asian students, both within Asia and as immigrants to Western countries such as Australia and New Zealand, to a culture where parents place high value on educational achievement and, as Gorur and Wu (2015) identify, to parents paying for private after-school tuition and cramming. Feniger and Lefstein (2014) also take a historical look and argue that this form of intense tuition and education "can be linked, at least in China, to preparations for Chinese civil service entrance examinations. This form of testing existed for about 1,300 years, from the seventh century until the early twentieth century" (p. 850). This resonates with the point made earlier for Finland – that past history echoes down the ages and influences education systems in the present. Furthermore, Eckert (2008), in identifying Japan as the top-performing country in the TIMSS video study, criticised the study for not including the fact that the majority of secondary students attend *juku* schools, private schools after a school day, and Eckert concludes that "the success of the Japanese students on the TIMSS does not necessarily correlate with the type of instruction provided just in the state schools in the study" (p. 205). Sjøberg and Jenkins (2020) also caution about taking PISA results at face value and give the example of Vietnam, where only 56% of 15-year-olds are in school; they argue therefore that "it is difficult to claim that Vietnamese schooling is a 'stunning success'" (p. 2). These findings suggest that a simple reading of PISA and TIMSS scores needs to be set alongside cultural and contextual factors that also influence test performance.

Finally, Gurur and Wu question the whole premise that reporting average scores in these international tests is a useful measure for understanding science achievement in Australia. In their paper, they break down the scores of students in different Australian states and compare these with China (Shanghai), a city, and other small countries such as Singapore, all of which have similar populations. Gorur and Wu (2015) found that:

> Comparing Australian jurisdiction results with those of the East Asian PISA elite, ACT (Australian Capitol Territory) is ranked fifth internationally by mean score. So, a part of Australia is already in the 'PISA top five'! Indeed, ACT's performance is not statistically significantly different from second-ranking Korea. Western Australia (WA) is ranked eighth internationally by mean score, and it is also not significantly different to Korea's performance.
>
> (p. 564)

They go on to argue that, with ACT and WA already performing in the top five of PISA, perhaps it is to these states that Australia should look to improve its performance, rather than drawing on practices from countries whose culture is different, and which may have very different systems of education, concluding that "given that education is deeply culturally embedded, practices and policies might not 'travel' that well across cultures. So it would make eminent sense to find within-country role models" (p. 654).

Conclusion

Addressing the three key questions has shown that understanding science pedagogies internationally and from a comparative perspective is complex. First, the finding that inquiry-based teaching is not associated with developing students' scientific literacy is far more complex than this headline finding suggests. As we have seen, findings from the secondary analyses of PISA and TIMSS data (see for example Oliver et al., 2019; Teig et al., 2018) have offered a more nuanced and fine-grained analysis and give useful directions for further research and investigation. They also provide food for thought for teacher educators in educating new and experienced teachers in inquiry-led teaching and the role of practical work, and the same questions also need to be asked of teacher directed methods and adaptive methods. Oliver et al. (2019) also make the same argument when they say:

> These finer-grained distinctions regarding the optimal frequencies at which various aspects of inquiry-based instruction in secondary school science become observably effective must also be important and of interest for teachers and teacher educators. For example, doing practical work in every lesson or very rarely is unlikely to support the development of students' scientific literacy. Importantly, these findings show that 'inquiry' is not only multi-faceted, but also its relationship with scientific literacy varies according to the particular strategy being examined, and is often best conceptualised as non-linear. An important message from this research is that teacher educators and policy makers hold responsibility to support the development of pre- and in-service teachers using carefully developed evidence that informs

recommended practice. In our view, this commitment to a finer-grained examination of pedagogical strategies applies as much to teacher-directed and adaptive instructional strategies as it does to inquiry-based teaching, as described here.

(p. 19)

Second, what have we then learned in relation to how culture influences science pedagogy internationally? Again, the picture is complex; for example, Lau and Lam's work suggested that there were no cultural clusters of distinct science practices internationally, whereas the TIMSS 1999 video study did identify certain characteristic practices at a national level. So, although there is no really definitive answer from these studies, this again suggests a direction for further research.

Finally, there is a question about whether anything can be learned from these higher-performing jurisdictions, given each country's unique culture and context. There are some powerful arguments that might suggest that this is indeed challenging. For example, the case of Australia suggested that it might be more beneficial for a country to look internally at its own high-performing states/ districts, which have similar cultures and contexts. It might also suggest a need, before a process of policy borrowing is undertaken, that some work be done to better understand the particular unique social, educational and cultural circumstances of a country.

As I have said, in addressing these questions there is no simple answer of what works best in science pedagogies internationally. The nature of this complexity is expressed well by Lau and Lam when they concluded that successful teaching is:

> an amalgam of didactic and constructivist pedagogy as a result of the Western progressive educational philosophy adapted to the Eastern sociocultural contexts, or the Confucian culture of diligence, high expectation and conformism adapted to the Western sociocultural contexts.

(p. 2129)

This suggests that many approaches can be appropriate depending on the particular context. Decisions about what is appropriate, I would argue, lie principally in the hands of teachers every day and in every classroom across the world. So, in order to embrace this complexity, perhaps a key priority is to focus on ensuring our teachers are adaptable teaching professionals, valued, trusted and with genuine professional autonomy. This would allow teachers to harness their adaptive expertise (see Berliner, 2004) in order to make decisions about which approach is appropriate to fashion a unique learning experience that is relevant, engaging, powerful and educative for all their learners and so that a compromise between achievement and engagement does not have to be made. This, of course, demands high levels of pre-service and ongoing in-service teacher education to educate teachers to develop high levels of adaptive expertise. But it may also require a significant change in attitudes about teachers in many jurisdictions. There are examples internationally where teachers are highly qualified and respected; for example, in the high-performing countries of Singapore and Finland. Both have a strong focus on recruiting highly qualified teachers and in both countries, teachers are trusted and held in the highest respect (Goodwin, 2012; Sahlberg, 2012). So, do we also need to turn our attention to looking at what we can learn about teacher education in high-performing countries and how countries have managed to create a highly qualified, respected and trusted profession, or would this policy borrowing suffer the same challenges reported earlier? Another chapter in itself!

Finally, alongside this, all teachers need to be able to have access "to carefully developed evidence that informs recommended practice" (Oliver et al., 2019, p. 19). Therefore, as a science education community, we need to continue to gather evidence through research. For example, exploring the issues raised in this chapter of why the amount of inquiry-based instruction seems to matter. What level of teacher guidance is optimal in inquiry-based teaching? What elements of inquiry-based instruction do genuinely promote student learning, and how can we enhance those which do not? What is the relationship between PISA science scores and measures of student attitudes and interest? What does effective teacher-directed instruction look like? How does a country's culture and history influence its teaching and learning? There are many more, of course, and the combination of research and continuing to use PISA and TIMSS data in a critical way, while also promoting quality teacher education, is crucial to ensure that science teaching and learning is high quality, culturally appropriate and engages all our young people.

Questions for further debate

1 When and how should an inquiry-based approach be used in science teaching?
2 How can trust in science teachers be improved?
3 What are the limits to science teacher autonomy and who should set them?

Suggested further reading

Crehan, L. (2016). *Cleverlands: The secrets behind the success of the world's education superpowers.* Unbound.
Whitty, G., & Furlong, J. (2017). *Knowledge and the study of education. An international exploration.* Oxford Studies in Comparative Education, Symposium Books.

References

Abrahams, I. (2009). Does practical work really motivate? A study of the affective value of practical work in secondary school science. *International Journal of Science Education, 31*(17), 2335–2353. https://doi.org/10.1080/09500690802342836
Abrahams, I. (2011). *Practical work in school science: A minds-on approach.* Continuum.

Abrahams, I., & Millar, R. (2008). Does practical work really work? A study of the effectiveness of practical work as a teaching and learning method in school science. *International Journal of Science Education, 30*(14), 1945–1969. https://doi.org/10.1080/09500690701749305

Alfieri, L., Brooks, P. J., Aldrich, N. J., & Tenenbaum, H. R. (2011). Does discovery-based instruction enhance learning? *Journal of Educational Psychology, 103*(1), 1–18.

Areepattamannil, S. (2012). Effects of inquiry-based science instruction on science achievement and interest in science: Evidence from Qatar. *The Journal of Educational Research, 105*(2), 134–146.

Baroutsis A., & Lingard, B. (2017). Counting and comparing school performance: An analysis of media coverage of PISA in Australia 2000–2014. *Journal of Education Policy, 32*(4), 432–449. https://doi.org/10.1080/02680939.2016.1252856

Berliner, D. C. (2004). Describing the behavior and documenting the accomplishments of expert teachers. *Bulletin of Science, Technology and Society, 24*(3), 200–212.

Drent, M., Meelissen, M. R. M., & van der Kleij, F. M. (2013). The contribution of TIMSS to the link between school and classroom factors and student achievement. *Journal of Curriculum Studies, 45*(2), 198–224.

Eckert, J. M. (2008). Trends in mathematics and science study (TIMSS): International accountability and implications for science instruction. *Research in Comparative and International Education, 3*(2), 202–210.

El Masri Y. H., Baird, J. A., & Graesser, A. (2016). Language effects in international testing: The case of PISA 2006 science items. *Assessment in Education: Principles, Policy & Practice, 23*(4), 427–455.

Feniger, Y., & Lefstein, A. (2014). How *not* to reason with PISA data: an ironic investigation. *Journal of Education Policy, 29*(6), 845–855.

Forbes, C. T., Neumann, K., & Schiepe-Tiska, A. (2020). Patterns of inquiry-based science instruction and student science achievement in PISA 2015. *International Journal of Science Education, 42*(5), 783–806. https://doi.org/10.1080/09500693.2020.1730017

Furtak, E. M., Seidel, T., Iverson, H., & Briggs, D. C. (2012). Experimental and quasi-experimental studies of inquiry-based science teaching. *Review of Educational Research, 82*(3), 300–329. https://doi.org/10.3102/0034654312457206

Gruber, K. H. (2006). The German 'PISA-Shock': some aspects of the extraordinary impact of the OECD's PISA study on the German education system. *Cross-national attraction in education: Accounts from England and Germany,* 195–208, Symposium Books.

Goodwin, A. L. (2012). Quality teachers, Singapore style. In L. Darling-Hammond & A. Lieberman (Eds.), *Teacher education around the world. Changing policies and practices* (pp. 22–43). Routledge.

Gorur, R., & Wu, M. (2015). Leaning too far? PISA, policy and Australia's 'top five' ambitions. *Discourse: Studies in the Cultural Politics of Education, 36*(5), 647–664. https://doi.org/10.1080/01596306.2014.930020

Harlen, W. (2013). *Assessment and inquiry-based science education: Issues in policy and practice.* Global Network of Science Academies (IAP) Science Education Programme.

Hmelo-Silver, C. E., Duncan, R. G., & Chinn, C. A. (2007). Scaffolding and achievement in problem-based and inquiry learning: A response to Kirschner, Sweller, and Clarke. *Educational Psychologist, 42*(2), 99–107.

Hopfenbeck, T. N., Lenkeit, J., El Masri, Y., Cantrell, K., Ryan, J., & Baird, J. A. (2018). Lessons learned from PISA: A systematic review of peer reviewed articles on the programme for international student assessment. *Scandinavian Journal of Educational Research, 62*(3), 333–353.

Jerrim, J., Oliver, M., & Sims, S. G. (2019). The relationship between inquiry-based teaching and students' achievement. New evidence from a longitudinal PISA study in England. *Learning and Instruction, 61*, 35–44.

Kirschner, P. A., Sweller, J., & Clark, R. E. (2006). Why minimal guidance during instruction does not work: An analysis of the failure of constructivist, discovery, problem-based, experiential, and inquiry-based teaching. *Educational Psychologist, 41*(2), 75–86.

Lau, K.-C., & Lam, T. Y.-P. (2017). Instructional practices and science performance of 10 top-performing regions in PISA 2015. *International Journal of Science Education, 39*(15), 2128–2149. https://doi.org/10.1080/09500693.2017.1387947

Lazonder, A., & Harmsen, R. (2016). Meta-analysis of inquiry-based learning. Effects of guidance. *Review of Educational Research, 86*(3), 681–718.

Lingard, B., & Grek, S. (2007). *The OECD, indicators and PISA: An exploration of events and theoretical perspectives. Fab-Q working paper 2.* University of Edinburgh. www.ces.ed.ac.uk/PDF%20Files/FabQ_WP2.pdf

McConney, A., Oliver, M. C., Woods-McConney, A., Schibeci, R., & Maor, D. (2014). Inquiry, engagement, and literacy in science: A retrospective, cross-national analysis using PISA 2006. *Science Education, 98*(6), 963–980. https://doi.org/10.1002/sce.21135

Minner, D. D., Levy, A. J., & Century, J. (2010). Inquiry-based science instruction—what is it and does it matter? Results from a research synthesis years 1984 to 2002. *Journal of Research in Science Teaching, 47*(4), 474–496. https://doi.org/10.1002/tea.20347

Nortvedt, G. A. (2018). Policy impact of PISA on mathematics education: The case of Norway. *European Journal of Psychology of Education, 33*(3), 427–444. http://dx.doi.org/10.1007/s10212-018-0378-9

OECD. (2007). *The programme for international student assessment: Executive summary* www.oecd.org/pisa/pisaproducts/39725224.pdf

OECD. (2016). *PISA 2015 results (volume I): Excellence and equity in education.* PISA, OECD Publishing. https://doi.org/10.1787/9789264266490-en

Oliver, M., McConney, A., & Woods-McConney, A. (2019). The efficacy of inquiry-based instruction in science: a comparative analysis of six countries using PISA 2015. *Research in Science Education*, 1–22.

Roth, K. J., Druker, S. L., Garnier, H. E., Lemmens, M., Chen, C., Kawanaka, T., & Gallimore, R. (2006). *Highlights from the TIMSS 1999 Video Study of eighth-grade science teaching* (NCES 2006–17). U. S. Department of Education, National Center for Education Statistics. U.S. Government Printing Office. https://nces.ed.gov/pubs2006/2006017.pdf

Sahlberg, P. (2011). PISA in Finland: An education miracle or an obstacle to change? *Centre for Education Policy Journal, 1*(3), 119–140.

Sahlberg, P. (2012). The most wanted: Teachers and teacher education in Finland. In L. Darling-Hammond & A. Lieberman (Eds.), *Teacher education around the world: Changing policies and practices* (pp. 1–21). Routledge.

Simola, H. (2005) The Finnish miracle of PISA: historical and sociological remarks on teaching and teacher education. *Comparative Education, 41*(4), 455–470. https://doi.org/10.1080/03050060500317810

Sjøberg, S. (2015). PISA and global educational governance – a critique of the project, its uses and implications. *Eurasia Journal of Mathematics, Science and Technology Education, 11*(1), 111–127.

Sjøberg, S. (2018). The power and paradoxes of PISA: Should inquiry-based science education be sacrificed to climb on the rankings? *Nordic Studies in Science Education, 14*(2), 186–202.

Sjøberg, S. (2019). The PISA-syndrome – How the OECD has hijacked the way we perceive pupils, schools and education. *Confero: Essays on Education, Philosophy and Politics, 7*(1), 12–65.

Sjøberg, S., & Jenkins, E. (2020). PISA: A political project and a research agenda. *Studies in Science Education*, 1–20. https://doi.org/10.1080/03057267.2020.1824473

Stockard, J., Wood, T. W., Coughlin, C., & Rasplica Khoury, C. (2018). The effectiveness of direct instruction curricula: A meta-analysis of a half century of research. *Review of Educational Research, 88*(4), 479–507.

Teig, N., Scherer, R., & Nilsen, T. (2018). More isn't always better: The curvilinear relationship between inquiry-based teaching and student achievement in science. *Learning and Instruction, 56*, 20–29.

Ustun, U., & Eryilmaz, A. (2018). Analysis of Finnish education system to question the reasons behind Finnish success in PISA. *Studies in Educational Research and Development, 2*, 93–114.

Waldow, F. (2009). What PISA did and did not do: Germany after the 'PISA shock'. *European Educational Research Journal, 8*(3), 476–483. https://doi.org/10.2304/eerj.2009.8.3.476

Zhao, Y. (2017). What works may hurt: Side effects in education. *Journal of Educational Change, 18*(1), 1–19. https://doi.org/10.1007/s10833-016-9294-4

Section 3

Debates about whole-school issues which have a science dimension

Chapter 12

Inclusion and equity in science education

Saima Salehjee

Introduction

In this chapter, I argue that while science teachers might willingly sign up to the United Nations' 2030 Sustainable Development Goals (SDG), which offer "inclusive and equitable quality education" (United Nations, 2015, p. 14) to all, in actuality they manage to do very little – not least because exclusion and inequity still exist in schools and science classrooms. Clearly, this is a debatable issue for teachers to question their understanding and self-perceptions of "inclusion" and "equity" and their accompanying approaches to promote quality education for all students. My primary audience here is science teachers, and this chapter will highlight what science teachers should be debating to develop inclusive and equitable teaching and learning practices. I will first draw on relevant official reports to illustrate the issues, followed by various teachers' debates as exemplars from UK schools.

In framing this debate, my focus will be based on SDG 4 – Quality Education – one of the 17 goals that have been proposed to transform education worldwide. SDG 4 aims to "ensure inclusive and equitable quality education and promote lifelong learning opportunities for all" (United Nations, 2015, p. 14). Sayed et al. (2018) argue that the term "inclusive quality education" in the report is "somewhat ambiguous and contested, reflecting both a narrow (disability) and broader (all forms of exclusion) focus" (p. 197). In addition to SDG 4, there are many other references to equity in the SDGs.

Placing the term "inclusive" alongside "equity" in offering quality education stresses an agenda for global social justice. UNESCO's intention is that "education is a fundamental human right and an enabling right", "education is a public good", and "gender equality is inextricably linked to the right to education for all" (2016, p. 8). In simple words, education is for everyone, and by extension, science education is for everyone. In this chapter, I challenge the idea that science education, as currently practised, really is for everyone. An agenda for quality science learning for all, and for social justice, is rooted in the provision of lifelong learning – regardless of people's age, ability or level of study. In all of this, the chapter aims to discuss what UK-based science teachers debate and

DOI: 10.4324/9781003137894-15

should be debating about their current understanding, perceptions and practices in promoting a lifelong learning mindset to promote quality education for all.

The results of the PISA tests in 2018 show that differences in attainment between countries are likely due to the differential impact of socioeconomic backgrounds (Schleicher, 2018). Similarly, the World Education Forum (WEF) (2015) acknowledges the differences between countries – however, it also highlights that such differences can be evident within countries. Teachers from developed countries also need to take exclusion and inequity seriously. However, prominent reports, such as that by the United Nations, have "a strong focus on developing countries, and relatively little on developed countries" (Vladimirova & Le Blanc, 2016, p. 270).

The Organisation for Economic Co-operation and Development (OECD, 2017) reviewed SDG 4 and indicated progress in underdeveloped countries to achieve this goal related to "school infrastructure and access to basic education. However, significant challenges remain for many countries in achieving targets that measure learning outcomes and equity" (p. 27). So, UK science teachers, for example, might view SDG 4 at a global level and feel satisfied that the UK or other European countries have greater equity, not least because they are not impoverished war-torn zones or areas of acute underdevelopment. As I have already mentioned, major policy reports seldom discuss needs for inclusive and equitable education within more economically developed countries, so it is possible that science teachers in the UK could find difficulty in fully appreciating this goal in their own circumstances. Of course, the issues of eradicating poverty, tackling climate change, promoting equity for all and providing access to quality education apply to all countries, not just the less economically developed ones. And so, a debate involving and allowing teachers to reflect on their practices and act on them to promote SDG 4 is critical.

Things are changing, however, and there is some evidence that this global agenda is being debated and incorporated within more economically developed countries. For example, in the UK, the All-Party Parliamentary Group on Diversity and Inclusion in STEM (APPG, 2020) discussed the specific need for inclusive and equitable quality education within science education. The APPG recognised that science education is for all and recommended the promotion of "fairness and equality, a greater diversity of thought and experience, and citizenship" (p. 12). This perspective can be understood with the help of Alberto Rodriguez's (2016) statement on equity and equality at an individual level:

> Equity refers to the enactment of specific policies and practices that ensure equitable access and opportunities for success for everyone. It is important to differentiate equity from equality. . . . In order to be equitable, we cannot treat everyone the same. To be equitable, we must treat individuals according to their needs and provide multiple opportunities for success.
>
> (p. 243)

So equity in science education, for Rodriguez, is not simply about fairness or offering equal treatment to all students or certain groups of students who share similar characteristics (such as socioeconomic status). It is also about providing multiple and varied support – at an individual level – according to specific (unique) needs and "celebrating that difference as a source of strength for the community at large" (p. 242). To focus on this kind of individual student support within a local UK context, I have chosen to highlight two issues that I think teachers might consider discussing as exemplars with their colleagues. In doing so, I incorporate the stories of four students (Benjamin, Celina, Danielle and Anisha) and outline their personalised needs for inclusive and equitable support and opportunities for success.

Providing inclusive and equitable support at a personal level

To identify the type of support required, Rodriguez's statement must be considered at a student's personal level. For example, Benjamin is British, Black, from a working-class background and has aspirations to become a physicist. In his own view, he sees himself to be privileged within British society in having access to quality science education, although he recognises that he has limited support from his parent, little access to employment networks (Patacchini & Zenou, 2012) and (very) limited financial support to fund higher education (Flap & Völker, 2008; Zuccotti, 2015). Having dark skin colour and masculine appearances, Benjamin might well be misrecognised by teachers as having low achievement in science and low aspirations of science post-16. Such misperceptions are not uncommon (Archer et al., 2015, 2020) and can often lead to racist biases, and to Benjamin himself feeling discrimination due to his obvious characteristics. This context eventually results in exclusion and inequity towards Benjamin – and others who share similar characteristics (e.g., Carlone et al., 2015). Accordingly, a group of science teachers can use Benjamin's example to evaluate Benjamin's teachers' practices by discussing:

1   Why and how misrecognition of a student, based on their visible characteristics, alters teachers' support being placed where it is not actually needed? and
2   What should be the next steps science teachers take to support their students' individual needs and/or ambitions of becoming scientists?

Benjamin's science teachers, from my standpoint, might need to understand that he does not necessarily require added inclusive and equitable provision – he is already highly motivated towards physics. Rather, he could need support to meet university physics students and lecturers, or physicists from the industry to help build his social networking. Benjamin needs help to access local authorities or financial advisors to support his aspirations for university study. So, learning from Benjamin's story, to promote inclusion and equity requires a change of mindset in grouping people based solely on visible features. To change our mentality as educators, teachers and society, we should make a regular habit of scrutinising our own perceptions, viewpoints, teaching and research practices, which dictate how people from a particular group should perform and behave; otherwise, we risk negative discrimination of a student's aspirations, self-preferences and choices based on their sex, the colour of their skin, nationality, religion, culture or heritage.

The second story is about Celina, aged 15, who migrated from Poland to England with her mother at the age of 10. Since she arrived in the UK, she has lived and studied in the schools located in a low socioeconomic area of East London. In her primary and early secondary schooling, she never wanted to continue with science education after the compulsory age of school science. This rejection of science education was not because Celina disliked school science or science teachers; instead, she didn't feel that science was for her mainly because of her migration status, working-class background and almost-failing science test results. But her disapproval towards the future uptake of science education changed over one year. Celina gives credit for this change to her Year 9 science teacher, who talked to all the students individually in the new academic year, including Celina, about their likes and dislikes about science. Celina told her teacher that she doesn't dislike science but will not continue with it as she "doesn't belong to science" as a migrant girl from a working-class Polish family. The next day, Celina's teacher gave her *Charlie and the Chocolate Factory* book and an animated DVD of *Inside Out*, where Celina's homework was to tease out science highlighted in the stories. The characters in the stories and a discussion with the teacher changed her viewpoint, as Celina stated: "[B]oth stories were about young people coping in a new setting like me (in a new country/city) where they felt excluded; still science was part of their lives and to mine".

Like Celina, I see that this change in her viewpoint was clearly initiated by her Year 9 science teacher, who used children's stories not only to talk about the science of the brain and memory (in the case of *Inside Out*) and the chemistry of chocolate (as the science theme of *Charlie and the Chocolate Factory*), but allowed Celina to discuss the social issues and so supported her to view that science is for everyone, and anyone can read, learn and use science despite their differential nationality, heritage or socioeconomic status. The science teacher didn't stop here; instead, they kept meeting Celina regularly for over a year in the school context but out of the science classroom, such as during break times, form tutor times, after-school homework clubs, etc. Moreover, in every meeting, the teacher gave Celina some tasks and discussed them, such as researching women scientists from working-class backgrounds, talking to Polish university students studying at the local university, exploring science and scientific practices in her out-of-school life and studying the role of contemporary and Polish Enlightenment Period science/scientists.

Unlike Benjamin's teachers, I see that Celina's science teacher first talked to Celina about her viewpoints, interests and perceptions and then gave her personalised support as tasks and discussions. Currently, Celina is looking into colleges to study Biology and Chemistry. However, it is too early to say whether or not she will continue with it at university. Still, it is for sure that Celina's teacher has successfully encouraged her not to discontinue science education after the compulsory age of science education. So, the debate here for science teachers keen to provide personalised support to their students could be:

1   Would intersecting science with social topics, using children's books, for example, help teachers to enquire about a diverse range of students' viewpoints, perceptions and interests about science?
2   What other approaches could teachers use to intersect science with students' personalised views of their own gender, ethnicity and social class?
3   Would aspirations into science in this way result in good exam results? Why or why not?

Providing inclusive and equitable support at a personal and 'intersectional' level

Most government initiatives dealing with equity tend to focus specifically on one marginalised element of identity. For example, the United Nations (2015) has pointed out the need to achieve "gender equality and empower all women and girls" (p. 14). Women in engineering interventions are widely adopted to reduce gender stereotyping (House of Commons, 2001), and the Institute of Physics has numerous initiatives to improve gender balance (Skills Development Scotland, 2018). However, focusing gender (Scantlebury, 2012), presents an incomplete picture of inequity and could be misleading. To achieve diversity at the level recommended by Rodriguez requires a greater level of reform (Bianchini, 2017), focusing on "multiple marginalized identities [that] are especially critical to advance equity" (Núñez et al., 2020, p. 97). Canfield et al. (2020), for example, demand an "intersectional approach" as a basis for this reform. The intersectional approach means addressing the relationships among diverse social categorisations (gender, ethnicity, social class, religion etc.) which are multiplicative (Jang, 2018).

Single issue-based government and institutional ambitions could result in teachers having "tunnel vision" (Crenshaw, 1989), leading to them developing exactly the same interventions for inspiring female engineering students, despite each girl's particular contexts and circumstances. For instance, in our 2015 study (Salehjee & Watts, 2015), Danielle, who is British, White and middle-class, viewed herself as incapable of doing science and felt, even from childhood, that she was never supposed to be a "science person". She was surrounded by a family of philosophers, poets and writers with privileged (her own word) access to books and theatre. In addition, she intersected this with her own perceptions of "being a woman" which, in her view, did not involve working in science laboratories – despite her having "great" science teachers at school. So, even while

Danielle's science teachers did a good job, she was a confirmed "science rejector" (Salehjee & Watts, 2015). This story highlights the viewpoint that, young women who share similar social class and family life to Danielle's cannot simply be aggregated into one group and then be seen to have the same meanings for science, life, preferences and actions.

A final example, Anisha, is British, South Asian and middle-class and, like Danielle, also opted away from science upon leaving school. Unlike Danielle, though, she viewed herself as a well-informed science person and mentioned that her mother wanted her to become a doctor, and that her own intention was to become a civil engineer. Still, Anisha discontinued science after school not because she did not "have it in her" to become a civil engineer, but, because of her immigrant background, this was not an option supported by her school science teacher. She remembered that:

> My science teacher – who happened to be my form tutor – told me that my grades in science are just average and . . . being new to the country would be difficult for me to study difficult subjects like science and maths.

Though both Danielle and Anisha opted out of science, their reasons for doing so were quite different, suggesting very different intersections of social structures and self-belief. Science teachers reading these two stories could consider debating about:

1 their experiences with students who disown science at a personal and intersectional level, and
2 what they have done/should have done to encourage students to appreciate science despite not being willing to take up science education and/or a career in the future.

Having highlighted the two debates on personalised teacher support at an intersectional level, I will continue with science classroom-based examples to embed inclusive and equitable teaching and learning approaches.

Establishing inclusive and equitable science classrooms to develop lifelong scientific learners

As mentioned earlier, UN SDG 4 emphasises the main purpose of inclusive and equitable quality education being to develop lifelong learners. This goal also highlights the need to provide scholarships to higher education and to vocational programmes for students from less economically developed countries in order to study in more economically developed countries (Webb et al., 2017). Closer to home, though, the UK's educational policies for lifelong learning in science are beset with problems (Salehjee & Watts, 2020). The key outcome of lifelong science learning in the UK context is not to make scientists, in my view, but to promote a "science responsibility" – or even a "science duty" – for all citizens (Owen et al., 2013; Roy et al., 2012) whether citizens are scientists, doctors or engineers (or not).

A range of recommendations have been made by policymakers who advocate vocational and higher education studies (United Nations, 2015) and work placements in STEM industries (Government Office for Science, 2017). These might be seen as a major "expansion of post-18 education and training to level up and prepare workers for the post-COVID economy" (Prime Minister's Office, 2020). These initiatives correctly emphasise – to varying degrees – the importance of focusing on lifelong learning initiatives with no end date to learning. There are no specific links made to the advancement of lifelong learning in England's school science curriculum (key stages 1–4). However, many studies in lifelong learning in science do promote scientific literacy, such as the European Parliament's Committee on Culture and Education (Siarova et al., 2019).

The Curriculum of Excellence (CfE) in Scotland uses the terms "lifelong" and "scientifically literate citizens" in the same sentence: "develop as a scientifically-literate citizen with a lifelong interest in the sciences" (p. 1). The CfE also supports Scottish students' future in sciences by developing "skills for learning, life and work" (p. 1), and links with aims for inclusion and equitable by providing opportunities to "express opinions and make decisions on social, moral, ethical, economic and environmental issues based upon sound understanding" (p. 1). One problem here, though, is that the start date of lifelong learning could be seen to come after the primary stage of science education. This gives the unfortunate message to early-years and primary teachers that they need not be concerned with developing students as lifelong science learners. The road to becoming a lifelong scientific learner, though, begins as early as foundation education (Marian & Jackson, 2017; Salehjee, 2020).

According to several authors, teaching for scientific literacy should begin very early (e.g., Kähler et al., 2020; Salehjee, 2020). Looking at England's Science Curriculum, for example, at key stage 4, the ideas that link to inclusive and equity learning are "everyday and technological applications", "social, economic and environmental implications" and "evaluating risks" within the "societal context" (p. 5). In key stages 1, 2 and 3, though, these aims for inclusive and equity-related science are much more muted. For instance, at key stage 3, the National Curriculum states that "social and economic implications of science are important but, generally, they are taught most appropriately within the wider school curriculum: teachers will wish to use different contexts to maximise their pupils' engagement with and motivation to study science" (p. 2). The wider school curriculum may well have this on its agenda, but the key stage science curriculum seems to be restricting teachers here.

Although there are good intentions in the science curriculum, the unit of work and associated topics in the science curriculum do not clearly define the term scientific literacy or align it with the development of scientifically literate people (e.g., Millar, 2008). I start this discussion by using Nancy Brickhouse's (2007) definition, and view scientifically literate people as those who can debate on inclusion/exclusion and equity/inequity (pertaining to social, moral, ethical, economic and environmental) issues of science by:

- demonstrating some level of scientific understanding;
- critiquing written/oral science-oriented texts and/or scientists for the betterment of public life;
- drawing upon scientific knowledge and skills to make personal choices for a better lifestyle, appreciate and understand scientific ideas concerning their interests; and
- critically analysing written, oral and/or visual texts to make scientifically informed decisions.

Aligning Brickhouse's four points with the school science curriculum, teachers might include some important aspects of exclusion and inequity debates – such as using texts from varying sources to debate and critique issues of pollution, the misuse of energy, inequitable differences in life expectancy between communities or resources and population issues (national and/or global). The science curriculum recommends that teachers make science relevant to students' everyday lives, in other words "personalising science", which, in turn, can also make it inclusive and equitable (Godec et al., 2017), though for some but not for all. For example, some students might find it difficult to engage in debates about water pollution when they have access to clean water, or to issues of poverty when their basic living needs are met – they might find it difficult to be concerned about overpopulation where their own family comprises only three members.

To make science learning inclusive and equitable, some science educators have recommended that teachers move away from localised "Eurocentric" teaching, for instance, by diversifying the geographical locations of science/scientists taught in the classrooms. Vongai Mpofu, Femi Otulaja and Emmanuel Mushayikwa (2014) recommended integrating African plant healing practices, using a combination of medicinal plants and spiritualism to manage health, as a way to incorporate indigenous knowledge in Western-oriented science classrooms. Some use the strand of "decolonising the (science) curriculum" in a secondary school context. Haira Gandolfi (2021), for example, has illustrated how secondary science teachers might decolonise England's science curriculum (DfE, 2015). Gandolfi incorporated topics on medicines, magnetism, evolution and Earth's resources, mainly dealing with building the geographical location and historical storyline in building in the aspects of social and cultural influences, issues of environmental and intellectual property of drugs, collaboration among different countries and some ethical controversies. Further, some people advocate interdisciplinarity by using the Critical Disciplinary Literacies (CDL) approach. CDL encourages teachers to use everyday (vernacular) language and associated knowledge to make use of multiple sources of texts (including pseudo-scientific texts), to embrace social functions and analyse them with social awareness (Race et al., 2022).

An opportunity for science teachers to debate here is, while these attempts to make science learning inclusive and equitable are commendable, do they capture the social issues involving gender, religion, nationality and race debates or not?

The teachers could self-reflect to see if this is the case in their teaching, too, as it is noted that the majority of science teachers shy away from these discussions (Banks et al., 2007; Race et al., 2022). Nevertheless, these issues are debated in society as elements that face discrimination or privilege in schools, universities and workplaces. Science teachers could argue against bringing these social debates into the science classrooms, perhaps because they are not relevant to the science they teach. Instead, it is the role of the social sciences teacher to include this perspective in their teaching. Christine Reich et al. (2010), though, argue that such classroom discussions can result in students challenging biases that promote, say, pseudo-scientific studies that root exclusion and inequity in society. Such discussions are the responsibility of both the science teachers *and* the social science teacher (like any other teacher).

Therefore, as a school including the science teachers, an ongoing debate is needed to recognise and act against exclusion and inequities towards marginalised individuals and communities. I illustrate some possible approaches here, focusing on science classrooms.

Gender debate: discussing science for science

In her book *Inferior*, Angela Saini (2017) discusses how science and scientists have contributed to creating inequity between male and female genders. Her contribution highlights the historical and contemporary views of science and scientists, which have shaped (and still shape) stereotypical societal viewpoints, for example that women are inferior to men. One of the many examples Saini used in her book includes the pseudo-scientific dichotomy of brain size, indicating that men are more intelligent as the size of a man's brain is larger than a woman's brain.

Sally, a student teacher in a secondary-based school in West London, wanted to bring Saini's example into the classroom and asked her mentor about the appropriateness of including it when teaching the central nervous system. The mentor replied "probably not", as "it would be time-consuming to debate unrelated conversations". Sally, with puzzlement, asked her mentor: "Why is it unrelated?" The mentor replied, "Because intelligence and feminism are not what the science curriculum asks for". This, though, is not the case – the science curriculum does have related topics and encourages teachers and learners to promote the scientific literacy elements of critique, analysis and evaluation. For example, a teacher can include questions to promote feminist pedagogy by creating a classroom student talk guiding the inquiry process environment (Mayberry, 2004, p. 212) such as: can MRI, CT or PET detect differences among male and female brains? Is there such a thing as a male or female brain? A science or maths brain? What differences do brains show during prenatal testing? How does the brain develop after childbirth, and how does it vary (or not) among male and female children?

I believe that a further discussion between Sally and her mentor would have mutually benefited both to see the relevance of gender and reveal scientific myths and stereotypical understandings and allow them to engage with dialogues on discrimination, privilege, lack of research and self-perceived stereotypical views. By discussing this gender inequity issue, Sally and her mentor could usefully elaborate their discussion, including conversations that women are the weaker sex, are less intelligent or incapable of achieving academic goals that men can achieve. Is Sally's mentor right in suggesting that these conversations will be time-consuming and irrelevant to the science curriculum? Here experienced teachers need to take a stand: are they supporting newly qualified teachers to be inclusive and equitable in their teaching practices, or just preparing them to teach the scientific content knowledge highlighted in the science curriculum? Or both?

Nationality debate: promoting global/national equity

James, a key stage 3 science teacher, and Martin, a key stage 3 and 4 geography teacher, decided to set an interdisciplinary (science and human geography) task for their 13-year-old students to complete during the Covid-19 national lockdown. Part of the task was to write an essay, using four news articles, on the environmental and societal impact of petroleum industries on the lives of the workers in those industries. James and Martin received students' reflective essays discussing the effects of life expectancy and associated health and safety regulations surrounding British Petroleum (BP). Both the teachers particularly appreciated those essays that addressed the impact at a global level by including debates on the inequitable global supply, energy consumption and the agenda of prioritising the planet over individual countries.

Such students' essay showcasing the civic and social dimensions of science, at a global level, according to James and Martin, open a horizon of knowledge and understanding of ethics and moral beliefs and promote the ability to critique inequalities induced by national and international policies and politics.

After completing this activity, James discussed with Martin that he would deepen the chemistry knowledge in the next lesson. James mentioned that he would first link the petroleum industries reflective essays with the teaching of fractional distillation of crude oil to the same class. James next planned to include a short video showing today's high-tech and health-and-safety-regulated petroleum industry governed by the UK. The video will then be followed by some student activities using Molymods (a molecular modelling kit) and a teacher demonstration exhibiting the distillation of crude oil based on different boiling points. Martin was quietly listening to James and agreeing to his plan up to now, but then he got puzzled with James' next step of including a video from Simon Revee's BBC2, "Around the World" documentary on Burma, showcasing an impoverished family living in Burma's oil-rich area who were unofficially drilling oil, wearing no personal protective equipment, were barefoot, and were exposed to fire risks and carcinogenic pollutants such as polycyclic aromatic hydrocarbons and heavy metals.

According to James, it will be an attempt to teach the chemistry of fractional distillation of crude oil by bringing in local and global contexts. But, as Martin sees it, James had unintentionally introduced the notion of British industry's superiority over Burmese. These superiority/inferiority aspects were presented, according to Martin, by showing in the first video that British companies are advanced and civilised. In contrast, the second video shows that Burma oil industries are illegal, but did not show or even mention Burma's legal petroleum industry, Burmah. However, Burmah's successor is none other than BP. So, Martin urged James to question before including the second video: how to overcome the inequity issue of superiority/inferiority, even if the teaching supports students' scientific knowledge and accommodates local and global contexts?

Religion debate: future scientific citizens

During a school science meeting, William – a principal teacher of science – began a conversation with his science colleagues, Claire and Josh, on how best to teach Charles Darwin's theory to his 14-year-old students. He started by saying, "Darwin's theory, including the evolution of humans from apes, is not widely accepted by monotheist's teachings. As Adam and Eve are viewed to be the first two people on Earth, so apes are not the ancestors of homo sapiens". Then he asked: "So should we allow students to talk freely and link their views on the evolution process of homo sapiens with (non)religion? Does it exclude/include the different religions of the students?" Claire replied, "It would be very risky to bring in students' religious beliefs – some might not have a religion or want to talk about it". Next, Josh said, "Maybe just ignore anti-religious aspects and teach what is ascribed in the science curriculum". So, the teachers decided to teach the topic – aligning mainly with Josh's recommendation of just teaching it as a topic without getting into (non)religious discussions. But they did manage to point out briefly that during Victorian times, Darwin's ideology was rejected on religious grounds and added an examination-style question in their lessons where Darwin was pictured as an ape.

At first, this seems to be a reasonable effort by William and his team, but then after teaching this topic, the three colleagues realise that their intended learning outcome was not as powerful as it could have been as they have ignored other existing (non)religious beliefs. Therefore, in the next meeting they agree that (non)religion-based discussion should be exposed – not only to cover Darwin's theory's scientific ideologies, but also to promote (in)equity debates on where and how science and religious beliefs coincide or are parallel with each other. This approach, they believe, would allow students to draw upon personal religious or non-religious beliefs supporting (or not) their self-perceived scientific knowledge and skills to make personal choices. However, Josh worry about how they can understand and embed diverse viewpoints from all the students in a classroom situation. Plus, they might find that attempting to promote equity could risk inequity-based discussions among students because of their own (the teachers') self-belief and some peers' authoritative viewpoints over others. In answering Josh, William said, "Well, teaching any controversial topic, such as ruling for and against abortion, a one-child policy or female genital mutilation ('cutting'), etc., would require us to take risks, but yes, it requires thorough planning and research about the topic". After some further discussion, the science team decided to first read about the different religious beliefs on evolution and talk to the school's religious education teacher to guide them to some resources available to the students. Second, minimise the risk of inequity-based discussions on religion, especially where specific religion-based topics could be uncomfortable for some or all students, by reflecting on questions such as: how they can appreciate all students' (non)religious viewpoints? How can they avoid colonial discussions among students on religion and science? How can they promote talk that accepts and respects students' different ideologies about their perception of scientific concepts and beliefs? How far should they go to embed such discussions? Finally, how to teach about the "scientific evolutionary account and a religious belief" as "a guiding creative force [which is] not just compatible, but mutually reinforcing" (Ward, 1996 quoted in Ruse, 2003, p. 368)?

After a few meetings, the team organised an hour-long school assembly and invited scholars to talk about their perspectives of evolution in the light of their practising faith, namely Atheism, Buddhism, Chirstianity, Hinduism, Islam and Judaism. Next, the teachers allowed students to converse about it in small groups, openly bring in their self-perspectives about evolution, and ask questions from the scholars. This discussion didn't stop after the school assembly; instead, it continued in the science classrooms where William, Claire and Josh felt confident and quite prepared to allow an open debate on intersecting the science of evolution with religious or non-religious beliefs.

Race debate: evaluating pseudo-science

John, a probationary science teacher, and his mother Aileen, a biology teacher for the last 30 years, are of African heritage. They have not missed a single episode of BBC1's 2021 *The Big Questions* series on moral, ethical and religious debates, hosted by Nicky Campbell. In one of the programmes, Kehinde Andrews was the invited guest, debating in light of his 2020 book *The New Age of Empire.*

John and Aileen found Andrews' viewpoints intriguing and bought his book to explore more. In his introduction, Andrews critiqued various Enlightenment thinkers and their pseudo-scientific beliefs, including Immanuel Kant. After just reading the first chapter, the mother and son unanimously believed that exploring the history of race in today's world is a useful activity that can help eradicate racism from society.

Their conversations continued and reached the point where Aileen said, "Then should we include Kant's pseudo-scientific approaches and their impact on Black and Brown people in our science classes?" One example is, "Race is hierarchical where Whites are at the top of civilisation, Blacks are at the bottom and Yellow in the middle". John replied, "Absolutely, I think it is important to talk about these historically rooted controversies as we, unfortunately, still witness them today". "Yes", said Aileen, and she started to mind-map the lesson plan: So, "In the next class – as planned – I will first allow the students to research the current statistics on the number of British African and Asian people deaths due to the Covid-19 pandemic. Next, compare and contrast their findings against predominantly African and Asian countries. Then in groups, ask them to plot the percentages in the line-graphs before asking them to discuss the big question: is it Covid-19 that is discriminating between different racial groups, or is the scientific research that is discriminating against them?" Following their discussion points, "I will bring Kant's pseudo-scientific viewpoints (discussed in Andrew's book) to show one of the root causes of race and racism in the society which is practised for many years".

I agree with Aileen and John that such crucial discussions need to be debated in the science classrooms. And the teachers should allow students to discuss race (or any other social factors) and its pseudo-scientific viewpoints by using their scientific knowledge to critique and use a moral lens to discuss such views. For example, debating about the advances in genetics research on DNA sequencing goes against the pseudo-scientific construct of DNA to prescribe race, which science today believes is a poor diversity marker.

Simultaneously, in moral terms, today's science critiques the hierarchical distribution of race, White supremacy and discusses the greater variety of diversity within the White, Black, Brown and Yellow groups of people than between them. The opportunity for teachers' debate could include how comfortable it would be for John and Aileen as British Black teachers, for example, to engage with predominantly White students or even to converse about their thoughts with other science colleagues. How comfortable would it be for a White teacher to talk about White supremacy? A discussion on these topics may be uncomfortable for many teachers and students. Still, they need to be discussed, and being White or non-White does not make one more or less capable of talking about race and racism. What it requires is that people (teachers here) go beyond their viewpoints of being non-racist to anti-racist (Andrews, 2021); for example, teachers can debate race and racism and how they, as inclusive and equitable science teachers, can action scientifically informed anti-racist teachings in their classrooms.

The enquiry project

Callum, the Faculty Head of Science in a Glasgow secondary school, planned an enquiry project along with his science department, including five science teachers, as part of the school's development plan to promote inclusion and equity in their teaching and learning approaches.

The project started with a short student questionnaire to collect students' perceptions about school science and how school science incorporates their self-belief on social issues of gender, nationality, religion, race and social class.

Forty percent of the students' responses indicated that they were unsure whether these social issues of gender, nationality, religion, race and social class intersected with what they learn in the science classrooms. Thirty-seven percent were confident that the science lessons do not involve such discussions. The remaining twenty three percent mentioned that these topics were discussed in some of their biology lessons, although very briefly.

This data initiated a critical discussion among the science teachers. After a three-month conversation, the teachers designed lessons embedding the social issues of gender, nationality, religion, race and social class as part of their enquiry project. These conversations included questions like:

(i) How to approach and represent inclusion and equity-related issues to the students concerning gender, nationality, religion, race and social class in all the lessons. What are the benefits and drawbacks of taking this approach? What risks are involved?

(ii) How best to incorporate the five social issues of gender, nationality, religion, and race, and intersect them with the science curriculum?

(iii) How to ensure that students' beliefs and perceptions are equally accepted and respected concerning gender, nationality, religion, race and social class?

Next, each teacher took one of the five social issues mentioned to plan lessons for the key stage 3 students and shared it with the other teachers. Before teaching these lessons, Callum organised a meeting with all the science team, and as a team, the teachers made the changes where required. The main points in these discussions were to ensure the:

1 relevance of the social issues with the science curriculum and students;
2 avoiding superiority and inferiority aspects;
3 giving equal opportunity to all the students to participate;
4 planning of time- and cost-effective resources.

This resulted in teachers teaching lessons that incorporated the five social issues to all the key stage 3 students. In addition, after teaching these lessons for over six months, questionnaires were given to the students to share their experiences of intersecting social issues with science and how these experiences have influenced their viewpoints about gender, nationality, religion, race and social class intersections with science.

Callum and his science team's enquiry project supported their aim of making inclusive and equitable lessons by focusing not only on one aspect of the social issue, but on a number of them. This project, in my view, exhibited three main benefits:

1 it allowed students' voices to be heard from the questionnaires;
2 teachers debated and scrutinised planning and corresponding teaching approaches to embed quality education;
3 teachers shared their resources and reduced the burden of time and workload.

Conclusion

This chapter has highlighted the following five points:

1 Quality education cannot be achieved without the immersion of inclusion and equity in the education system.
2 The social justice agenda of inclusion and equity varies from global North to global South countries, but it is equally essential for all nations to maintain quality education.

3   Promoting support at an individual and intersectional level requires teachers to change their mindsets of grouping people's ability to do science on their visible features and regularly reflect on their assumptions and actions dictating how a particular group of people should behave and perform.

4   It is the teacher's responsibility, including science teachers and not only the social science teachers, to include social issues based on inclusion/exclusion and equity/inequity relationships.

5   To promote a lifelong science learning mindset, teachers work as a team to promote the intersection of scientific learning with the issues that make science unattractive for women, immigrants, migrants, working-class and/or non-White British people.

Questions for further debate

1   How could teachers make science learning inclusive and equitable?
2   How might teachers be encouraged to adopt more inclusive and equitable approaches?

Suggested further reading

Gouvea, J. S. (2018). Culture and equity in science classrooms. *CBE – Life Sciences Education*, *17*(4), fe8. https://doi.org/10.1187/cbe.18–07–0124
Saini, A. (2019). *Superior: The return of race science*. Beacon Press.
Salehjee, S., & Catriona Cunningham. (2021). *Developing an anti-racist approach to teaching. Anti-Racist Curriculum (ARC) project*. Advance HE, QAA, Enhancement Themes and Scottish Funding Council. www.advance-he.ac.uk/sites/default/files/2021-10/5%20Developing%20an%20anti-racist%20approach%20to%20teaching.pdf

References

All-Party Parliamentary Group (APPG). (2020). *Inquiry on equity in STEM education.* www.britishscienceassociation.org/Handlers/Download.ashx?IDMF=debdf2fb-5e80-48ce-b8e5-53aa8b09cccc
Andrews, K. (2021). *The new age of empire: how racism and colonialism still rule the world*. Penguin UK.
Archer, L., Dewitt, J., & Osborne, J. (2015). Is science for us? Black students' and parents' views of science and science careers. *Science Education*, *99*(2), 199–237.
Archer, L., Moote, J., Macleod, E., Francis, B., & DeWitt, J. (2020). *ASPIRES 2: Young people's science and career aspirations, age 10–19*. https://discovery.ucl.ac.uk/id/eprint/10092041/15/Moote_9538%20UCL%20Aspires%202%20report%20full%20online%20version.pdf
Banks, J. A., Au, K. H., Ball, A. F., Bell, P., Gordon, E. W., Gutiérrez, K. D., Brice Heath, S., Lee, C. D., Lee, Y., Mahiri, J., Nasir, N. S., Valdés, G., & Zhou, M. (2007). Learning in and out of school in diverse environments. In *The LIFE center and center for multicultural education*. University of Washington.

Bianchini, J. A. (2017). Equity in science education. In K. Taber & B. Akpan (Eds.), *Science education* (pp. 453–464). Brill Sense.

Brickhouse, N. W. (2007). Scientific literates: What do they do? Who are they? *Proceedings of the Linnaeus Tercentenary 2007 Symposium Promoting Scientific Literacy: Science Education Research in Transaction 28–29 May 2007, Uppsala, Sweden.*

Canfield, K. N., Menezes, S., Matsuda, S. B., Moore, A., Mosley Austin, A. N., Dewsbury, B., Feliú-Mójer, M. I., McDuffie, K. W. B., Moore, K., Reich, C. A., Smith, H. M., & Taylor, C. (2020). Science communication demands a critical approach that centers inclusion, equity, and intersectionality. *Frontiers in Communication, 5*, 2. www.frontiersin.org/articles/10.3389/fcomm.2020.00002/full

Carlone, H. B., Johnson, A., & Scott, C. M. (2015). Agency amidst formidable structures: How girls perform gender in science class. *Journal of Research in Science Teaching, 52*(4), 474–488.

Crenshaw, K. (1989). Demarginalizing the intersection of race and sex: A black feminist critique of antidiscrimination doctrine, feminist theory, and antiracist politics. *University of Chicago Legal Forum,* (1), 139–167.

Department for Education (DfE). (2015). *National curriculum in England: Science programmes of study.* Author. https://dera.ioe.ac.uk/22953/1/National%20curriculum%20in%20England%20science%20programmes%20of%20study%20-%20GOV_UK.pdf

Flap, H., & Völker, B. (2008). Social, cultural, and economic capital and job attainment: The position generator as a measure of cultural and economic resources. *Social Capital: An International Research Program,* 65–80.

Godec, S., King, H., & Archer, L. (2017). *The science capital teaching approach: Engaging students with science, promoting social justice.* University College London.

Government Office for Science. (2017). *Future of skills & lifelong learning.* https://assets.publishing.service.gov.uk/government/uploads/system/uploads/attachment_data/file/727776/Foresight-future-of-skills-lifelong-learning_V8.pdf

House of Commons. (2001). *House of Commons science and technology-sixth report.* https://publications.parliament.uk/pa/cm200001/cmselect/cmsctech/200/20007.htm

Jang, S. T. (2018). The implications of intersectionality on Southeast Asian female students' educational outcomes in the United States: A critical quantitative intersectionality analysis. *American Educational Research Journal, 55*(6), 1268–1306.

Kähler, J., Hahn, I., & Köller, O. (2020). The development of early scientific literacy gaps in kindergarten children. *International Journal of Science Education, 42*(12), 1988–2007. https://doi.org/10.1080/09500693.2020.1808908

Marian, H., & Jackson, C. (2017). Inquiry-based learning: A framework for assessing science in the early years. *Early Child Development and Care, 187,* 221–232. https://doi.org/10.1080/03004430.2016.1237563

Mayberry, M. (2004). Connecting girls and science: Constructivism, feminism, and science education reform. *NWSA Journal, 16*(2), 212–214.

Millar, R. (2008). Taking scientific literacy seriously as a curriculum aim. In *Asia-Pacific Forum on Science Learning and Teaching, 9*(2), 1–18.

Mpofu, V., Otulaja, F. S., & Mushayikwa, E. (2014). Towards culturally relevant classroom science: A theoretical framework focusing on traditional plant healing. *Cultural Studies of Science Education, 9*(1), 221–242.

Núñez, A. M., Rivera, J., & Hallmark, T. (2020). Applying an intersectionality lens to expand equity in the geosciences. *Journal of Geoscience Education*, *68*(2), 97–114.

OECD. (2017). *Education at a glance 2017*. OECD Publishing. https://doi.org/10.1787/eag-2017-6-en

Owen, R., Bessant, J. R., & Heintz, M. (Eds.). (2013). *Responsible innovation: Managing the responsible emergence of science and innovation in society*. John Wiley & Sons.

Patacchini, E., & Zenou, Y. (2012). Ethnic networks and employment outcomes. *Regional Science and Urban Economics*, *42*(6), 938–949.

Prime Minister's Office. (2020). *Major expansion of post-18 education and training to level up and prepare workers for post-Covid economy*. www.gov.uk/government/news/major-expansion-of-post-18-education-and-training-to-level-up-and-prepare-workers-for-post-covid-economy

Race, R., Ayling, P., Boath, L., Chetty, D., Hassan, N., Mckinney, S., Riaz, N., & Salehjee, S. (2022). *Proclamations and provocations. Decolonising curriculum in education research and professional practice*. UCL Press.

Reich, C., Price, J., Rubin, E., & Steiner, M. (2010). *Inclusion, disabilities, and informal science learning*. A CAISE Inquiry Group Report. Center for Advancement of Informal Science Education (CAISE).

Rodriguez, A. J. (2016). For whom do we do equity and social justice work? Recasting the discourse about the other to effect transformative change. In N. M. Joseph, C. Haynes, & F. Cobb (Eds.), *Interrogating whiteness and relinquishing power: White faculty's commitment to racial consciousness in STEM classrooms* (pp. 241–252). Peter Lang.

Roy, H. E., Pocock, M. J. O., Preston, C. D., Roy, D. B., Savage, J., Tweddle, J. C., & Robinson, L. D. (2012). *Understanding citizen science & environmental monitoring. Final report on behalf of UK-EOF*. NERC Centre for Ecology & Hydrology and Natural History Museum.

Ruse, M. (2003). Belief in God in a Darwinian age. In J. Hodge & G. Radick (Eds.), *The Cambridge companion to Darwin* (pp. 368–389). Cambridge University Press. http://dx.doi.org/10.1017/CCOL9780521884754.016

Saini, A. (2017). *Inferior: How science got women wrong and the new research that's rewriting the story*. Beacon Press.

Salehjee, S. (2020). Teaching science through stories: mounting scientific enquiry. *Early Child Development and Care*, *190*(1), 79–90. https://doi.org/10.1080/03004430.2019.1653554

Salehjee, S., & Watts, M. (2015). Science lives: School choices and 'natural tendencies'. *International Journal of Science Education*, *37*(4), 727–743.

Salehjee, S., & Watts, M. (2020). *Becoming scientific: Developing science across the life-course*. Cambridge Scholars Publishing.

Sayed, Y., Ahmed, R., & Mogliacci, R. (2018). The 2030 global education agenda and the SDGs: Process, policy and prospects. *Global Education Policy and International Development: New Agendas, Issues and Policies*, *40*, 185.

Scantlebury, K. (2012). Still part of the conversation: Gender issues in science education. In *Second international handbook of science education* (pp. 499–512). Springer.

Schleicher, A. (2018). *PISA 2018: Insights and interpretations*. www.oecd.org/pisa/PISA%202018%20Insights%20and%20Interpretations%20FINAL%20PDF.pdf

Siarova, H., Sternadel, D. & Szőnyi, E. (2019). *Research for CULT committee – science and scientific literacy as an educational challenge.* The European Parliament's Committee on Culture and Education. www.europarl.europa.eu/RegData/etudes/STUD/2019/629188/IPOL_STU(2019)629188_EN.pdf

Skills Development Scotland. (2018). *Review of improving gender balance Scotland.* https://www.skillsdevelopmentscotland.co.uk/media/44705/review-of-improving-gender-balance-2018.pdf

UNESCO. (2016). *Unpacking sustainable development goal 4 education 2030 guide.* http://unesdoc.unesco.org/images/0024/002463/246300E.pdf

United Nations. (2015). *Transforming our world: The 2030 agenda for sustainable development.* https://sustainabledevelopment.un.org/content/documents/21252030%20Agenda%20for%20Sustainable%20Development%20web.pdf

Vladimirova, K., & Le Blanc, D. (2016). Exploring links between education and sustainable development goals through the lens of UN flagship reports. *Sustainable Development, 24*(4), 254–271.

Ward, K. (1996). *Religion and creation.* Clarendon Press.

Webb, S., Holford, J., Hodge, S., Milana, M., & Waller, R. (2017). Lifelong learning for quality education: Exploring the neglected aspect of sustainable development goal 4. *International Journal of Lifelong Education, 36*(5), 509–511. https://doi.org/10.1080/02601370.2017.1398489

World Education Forum (WEF). (2015). *Education 2030, Incheon declaration and framework for action: Towards inclusive and equitable quality education and lifelong learning for all.* ED-2016/WS/2. Unesco.

Zuccotti, C. V. (2015). Do parents matter? Revisiting ethnic penalties in occupation among second generation ethnic minorities in England and Wales. *Sociology, 49*(2), 229–251.

Chapter 13

Faith, science and classrooms

Michael J. Reiss

Introduction

The importance of religion varies greatly between people (all absorbing, of no interest, abhorrent) and between countries (contrast theocracies and atheist/secular jurisdictions), but is difficult to measure. Surveys may attempt to determine religious beliefs (e.g., faith in God), religious practices (e.g., how often someone goes to a communal place of worship or reads the scriptures of their religion) and religious experiences (e.g., feeling that God is talking to one), and these three aspects of religion are sometimes aggregated into 'religiosity', which can be characterised as 'religious commitment'. However, there is considerable disagreement as to how best to ascertain each of these, with religious practice probably being the most objective, and some of the terms used in surveys and interviews don't translate very well between religions and languages.

Perhaps the best international data on religious commitment are assembled by the Pew Research Center. In their most recent publication, weekly worship varies from 1% of the population in China, 6% in Sweden, 7% in Russia and 8% in the UK to 89% in Nigeria, 82% in Ethiopia and 72% in Indonesia, while the percentage who say that religion is very important in their lives varies from 3% in China, 10% in Japan, 10% in Sweden and 10% in the UK to 98% in Ethiopia, 94% in Pakistan and 93% in Indonesia (Pew Research Center, 2018). For a teacher, whatever their subject, what this means is that in most countries in the world, classrooms are going to contain some students for whom religion is important and some for whom it is not.

Many science teachers might presume that there is no place for religion in the science classroom (Reiss, 2008). Indeed, although I am unaware of any international analysis, it seems that religion features only infrequently in school science curricula, and then either only in the context of specific topics (e.g., evolution) or through general statements, such as the requirement in England for the curriculum to provide for spiritual, moral, social and cultural development. Apropos of this, the 2019 Ofsted School Inspection handbook states that:

DOI: 10.4324/9781003137894-16

The spiritual development of pupils is shown by their:

- ability to be reflective about their own beliefs (religious or otherwise) and perspective on life
- knowledge of, and respect for, different people's faiths, feelings and values
- sense of enjoyment and fascination in learning about themselves, others and the world around them
- use of imagination and creativity in their learning
- willingness to reflect on their experiences

(Ofsted, 2019)

The chief arguments, it seems to me, for school science teaching taking account of religion are threefold: religion is important to some students and, as discussed later, there are points where science and religion do intersect; there may be a curriculum requirement; good teaching about religion can enhance the quality of science teaching. The third of these arguments may be doubted by some but, if correct, it seems to be a particularly strong reason for science teachers, whatever their own beliefs and views about religion, to include some teaching about religion in science lessons.

Faith is generally taken as referring to religion or its more diffuse counterpart, spirituality. However, there has long been discussion within the study of religions as to what is included within 'religion'. For a start, there are arguments whether Buddhism is a religion or not. Although generally treated as such, Buddhism does not believe in a personal god, nor does it entail the worship of any supernatural entities. Then, there has long been a debate as to whether Marxism is a religion (e.g., Ling, 1980), while some have argued that nationalism can be considered a religion (cf. Mentzel, 2020). A different approach to broadening what is included within 'religion' is to adopt a term like 'religions and non-religious worldviews'. I will return to worldviews later; here, I simply note that while contentious (e.g., Barnes, 2020), the inclusion of worldviews alongside religion has been welcomed by many humanists and some religious educators (e.g., Commission on Religious Education, 2018).

Nature of science

The nature of science is taught in school science in many countries, and there is a large science education literature as to what is meant by the nature of science and how it might be taught. It is possible that some students might better appreciate how science builds up scientific knowledge if they compared it to how other disciplines understand the nature of reliable knowledge (e.g., Stones et al., 2020; Pearce et al., 2021). Consider the question of 'authority' and the scriptures as a source of authority. To the great majority of religious believers, the scriptures of their religion (the Tanakh, the Christian Bible, the Qur'an, the Vedas, the Guru Granth Sahib, the various collections in Buddhism, etc.) have authority by very virtue of their being scripture. This is completely different from the authority of science (Reiss, 2018a). Isaac Newton's *Principia* and Darwin's *On the Origin of Species* are wonderful books, but they do not have any authority other than that which derives from their temporary (albeit long-lasting) success in explaining phenomena of the natural world. All scientific writing, however brilliant, is eventually replaced. For example, as is well-known, Darwin knew almost nothing of the mechanism of inheritance despite the whole of his argument relying on inheritance, so parts of *The Origin* were completely out of date over a hundred years ago, once Mendelian genetics become recognised, understood and accepted.

More generally, students might be helped by their teachers to think about how science differs not only from religion but from other disciplines – such as history, mathematics and the arts. This is not to imply that there are always clear lines of demarcation; indeed, some of the times when science and religion are perceived to be in conflict is precisely when they make claims about the same areas. Consider, for example, the possibility of miracles, where the word is used not in its everyday sense (and the sense in which it is sometimes used in scripture), namely 'completely unexpected' or 'marvellous' (as in newspaper accounts when someone 'miraculously' finds the wedding ring they lost years earlier), but in its more precise meaning of 'contrary to the laws of nature'. Many religions are rather keen on miracles; such miracles indicate God's authority and power over the whole of creation; they also reveal something of God's nature and intent. Scientists, though, whatever their personal beliefs, simply do not permit miracles as explanatory possibilities in their work. As scientists, they are likely to react to reports of miracles in one of three ways: (i) miracles are impossible (because they are contrary to the laws of nature); (ii) miracles are outside of science (because they are contrary to the laws of nature); for example, the question of whether someone can be resurrected from the dead is simply not amenable to scientific inquiry; or (iii) miracles are very rare events that haven't yet been incorporated within the body of science, but will be (as certain rare events that were once considered remarkable, even miraculous, such as eclipses and occasional farm animals born with two heads or seven legs, have been).

Pedagogical approaches

There are a range of pedagogical approaches that might be used to address the science-religion issue in school science classrooms, whatever the age of the students (e.g., Owens et al., 2018; Billingsley et al., 2019). There isn't yet a consensus as to which of these is best. It might be that different approaches work better for different teachers, for different students (e.g., depending on their age) and for different topics.

Controversial issues

Issues can be controversial for a variety of reasons, whether they are issues that might be considered in science lessons (Levinson, 2006) or more generally.

At its simplest, a controversial issue is one where people hold different views about which they believe strongly. A useful distinction in science education is between issues where the controversy is because the science is uncertain, and issues where the controversy is for other reasons. For example, evolution can be considered a controversial issue, but the vast majority of scientists would hold that this is not because the science is uncertain, but because the well-established conclusions of science conflict with how some religious believers understand their scriptures.

There are several ways in which a science teacher might handle controversial issues in the classroom (Reiss, 1993). One approach is that of *advocacy*. Here, the teacher argues for a particular position. Suppose, for example, that a teacher strongly believes that the right thing to do is to ensure that countries move rapidly towards sustainable energy production. Such a teacher might advocate cutting back on fossil fuels and expanding the use of solar power, wind power, tidal power, biofuels, nuclear power and geothermal power. Another teacher might also believe that the right thing to do is to ensure that countries move rapidly towards sustainable energy production, but advocate that we do not increase our use of nuclear power on the grounds that it generates too much waste and occasionally leads to disasters (Chernobyl, Fukushima). A third teacher might advocate that we need a mixed economy of power generation on the grounds that renewables on their own cannot ensure a steady supply of electricity. The very fact that one can envisage this range of arguments indicates a weakness with the approach of advocacy – it is likely that a teacher will be able to outargue their students, whereas what we presumably want is not for students to be browbeaten into accepting their teacher's point of view, but to develop their knowledge and skills to consider and evaluate a range of possibilities.

Other approaches seek some sort of balance. In the approach of *affirmative neutrality*, the teacher presents different sides of a controversy, without indicating their own views. In the approach of *procedural neutrality*, the teacher elicits information about the controversy and different points of view from students, using materials that provide a range of viewpoints. Again, the teacher does not reveal their own position. However, Oulton et al. (2004) are sceptical about the possibility of balance, arguing that instead it is better to develop skills of criticality in students so that they become able to detect bias.

Worldviews

There is a huge science education literature on the misconceptions that students often hold about scientific topics, but over the last decade or so, there has been growing interest in the approach of worldviews as a way of understanding why students may not only not accept scientific conceptions, but also be resistant to changing their minds. Bill Cobern (1996) argued that science educators generally presume that if students can see that scientifically orthodox conceptions are more intelligible, plausible and fruitful than other conceptions, they will come to accept these scientific conceptions. Cobern characterised this as a 'rationalistic view'. He went on to point out that the notion of worldviews provides a different way of understanding why people hold the views that they do.

A worldview is a fairly coherent way of understanding ('viewing') reality ('the world'). I write 'fairly' because the criterion of coherence can be less important to many people than it is to a mathematician or a scientist, for example, for whom internal contradictions in a theory are generally fatal to that theory. However, suppose I have (for whatever reason) a worldview informed by neoliberal capitalism that sees competition as being beneficial not only in a field like retail, but also in education and healthcare. In that case, even considerable data on the failings of neoliberalism in education and healthcare would probably be insufficient to cause me to change my mind. It might be that however much data there was, I would never change my mind – rather as some parents continue to believe that a child of theirs is gifted and charming, despite few others agreeing – or that if I do change my mind, I do so with great reluctance and many caveats, still keen to give neoliberalism the benefit of the doubt.

One way that I have found can be helpful for students to deepen their understanding of the science-religion issue is to get them explicitly to think about the relationship between scientific knowledge and religious knowledge (Reiss, 2008). One can ask students, either on their own or in pairs, to illustrate this by means of a drawing, and then everyone in the class can discuss the various drawings that result. See, for example, the hypothetical representation in Figure 13.1. A person producing what is drawn in Figure 13.1 sees both religious and scientific knowledge as existing but envisages the scope of religious knowledge as being smaller than that of scientific knowledge and of there being no overlap between the two.

There are others for whom scientific knowledge and religious knowledge are not distinct. Close to one end are those who draw religious knowledge as being

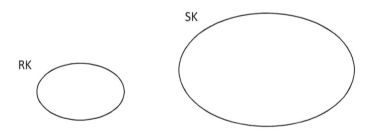

Figure 13.1 Hypothetical representation of how someone who sees both religious and scientific knowledge as existing but envisages the scope of religious knowledge as being smaller than that of scientific knowledge and of there being no overlap between the two might draw the relationship between religious knowledge (RK, left) and scientific knowledge (SK, right)

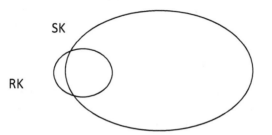

Figure 13.2 Hypothetical representation of how someone who sees religious knowledge as being much smaller than scientific knowledge and almost entirely contained within it might draw the relationship between religious knowledge (RK) and scientific knowledge (SK)

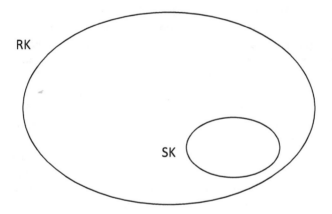

Figure 13.3 Hypothetical representation of how someone whose worldview is predominantly religious might draw the relationship between religious knowledge (RK) and scientific knowledge (SK)

much smaller than scientific knowledge, and wholly or partly contained within it (Figure 13.2); at the other are those whose worldview is predominantly religious (Figure 13.3).

Sensitive issues

More recently, I have suggested that there may be a benefit to treating certain *causes célèbres* in the science-religion field, notably evolution, as sensitive issues (Reiss, 2019). As humans, and this is probably particularly the case for teachers, most of us are quite good at knowing how to behave when we find ourselves with someone for whom an issue is sensitive (think a bereaved friend or colleague, or someone worried about their sexual identity or whether their country should go to war): we empathise, we are careful with our language, more tentative in our assertions, more alert to the possibility that the other person may be upset by something we say or some feature of our non-verbal communication, more tolerant of them. This is not quite the same as dealing with an issue that is controversial by virtue of people holding different opinions about it. In discussing such a controversial issue we may, depending on our personality, set out to win an argument – to convince the other person (our 'opponent') to change their mind, to come around to our point of view.

Socio-scientific issues

A final possibility is to see many of the intersections between science and religion as being somewhat akin to socio-scientific issues. There has been an explosion of writing about socio-scientific issues in the science classroom over the last 20 years (e.g., Kolstø, 2001; Zeidler, 2003; Sadler, 2009), although I am unaware of suggestions that it might be profitable to consider science-religion issues in this way. The central point of the socio-scientific movement is that it accepts two things: first, that many of the issues that engage learners and require an understanding of science draw on other disciplines too; second, that understanding these issues cannot therefore be reduced to a scientific understanding.

Consider, for example, the question of genetically modified foods. This issue is included in many school science courses, and it is common, after time on the science (DNA, protein synthesis) and technology (moving genes between species, gene editing) involved, to move on to ethical, economic and political issues. Students typically discuss reasons why such foods might be desirable (e.g., increased crop yields, longer shelf-life, more nutritious foods like so-called golden rice, cheaper foods, more choice for consumers) or not (risks of 'super weeds', greater use of pesticides, greater involvement of big business leading to traditional farming practices being crowded out, concerns that such foods are 'unnatural'). Some of the arguments are to do with religion – for example, whether such technologies entail 'playing God' and, if so, whether this is encouraged or frowned upon, whether 'patenting life' is acceptable – and one advantage of the socio-scientific issues approach is that no great theological expertise is expected of science teachers, just as they aren't expected to be experts in ethics, economics and politics.

Particular scientific topics

Evolution

Of all the science-religion issues that feature in school science, evolution is perhaps the most notorious. Many teachers of biology will have found that some students, or their parents, do not accept the theory of evolution and may even object to how the topic is taught in school.

Creationism exists in a number of different versions but, depending on the country, from as few as 5% to over 50% of adults reject the theory of evolution. Instead, they believe that life came into existence as described by a literal (fundamentalist) reading of the early parts of the Bible, the Qu'ran or other scriptures, and that the most that evolution has done is to change species into related species (Miller et al., 2006; Reiss, 2018b). For a creationist it is possible, for example, that the various species of hedgehogs (17 are generally identified, classified into five genera) had a common ancestor, but this is not the case for the various species of hedgehogs, deer and cats – still less for monkeys and humans, for birds and reptiles or for whales and soft fruit.

Allied to creationism is the theory of intelligent design. While many of those who advocate intelligent design have been involved in the creationism movement, to the extent that the US courts have determined that the country's First Amendment separation of religion and the State precludes the teaching of intelligent design in public schools (Moore, 2007), intelligent design can claim to be a theory that simply critiques evolutionary biology rather than advocating for or requiring religious faith. Those who promote intelligent design often make no reference to the scriptures or a deity, simply arguing that the intricacy of what we see in the natural world, including at a sub-cellular level, provides strong evidence for the existence of an intelligence behind this (e.g., Behe, 1996; Meyer, 2010). Natural selection, acting on its own from inorganic precursors, is held to be utterly inadequate for the diversity of life in all its richness that we see around us.

The scientific understanding of biodiversity is far from complete, but it holds that by almost 4 billion years ago, life had evolved on Earth. By the time of the earliest fossils, life was unicellular and bacteria-like. Gradually, the workings of evolution eventually resulted in the more than five million species, including our own, that exist today.

Whereas there is only one mainstream scientific understanding of biodiversity (the theory of evolution), there are a number of religious ones. Many religious believers are perfectly comfortable with the scientific understanding, either on its own or accompanied by a belief that evolution in some sense takes place within God's overall control, whether or not God is presumed to have intervened or acted providentially at certain key points, e.g., the origin of life or the evolution of humans – indeed, such positions are quite widely held (Laats, 2021). But many other religious believers adopt a more creationist perspective or that of intelligent design.

Most of what is written about creationism (and/or intelligent design) and evolutionary theory puts them in stark opposition. Evolution is consistently presented in creationist books and articles as illogical (e.g., natural selection cannot, on account of the second law of thermodynamics, create order out of disorder; mutations are always deleterious and so cannot lead to adaptations), contradicted by the scientific evidence (e.g., the fossil record shows human footprints alongside animals, such as dinosaurs, supposed by evolutionists to be long extinct; the fossil record does not provide evidence for transitional forms), the product of non-scientific reasoning (e.g., the early history of life would require life to arise from inorganic matter – a form of spontaneous generation long since rejected by scientists), the product of those who ridicule the word of God, and a cause of a whole range of social evils (from eugenics, Marxism, Nazism and racism to juvenile delinquency).

Creationism typically receives similarly short shrift from those who accept the theory of evolution. In a fairly early study, the philosopher of science Philip Kitcher argued, 'In attacking the methods of evolutionary biology, Creationists are actually criticizing methods that are used throughout science' (Kitcher, 1983, pp. 4–5). Kitcher concluded that the flat-Earth theory, the chemistry of the four elements, and mediaeval astrology 'have just as much claim to rival current scientific views as Creationism does to challenge evolutionary biology' (p. 5). The biologist Jerry Coyne puts it considerably less politely (Coyne, 2009).

Given all this, how might a science teacher teach about the theory of evolution if there are students who reject it on religious grounds? There is much to be said for allowing students to raise any doubts they have about the science of evolution and doing one's best as a teacher in such circumstances to have a genuine *scientific* discussion about the issues raised. The word 'genuine' does not mean that creationism or intelligent design deserve equal time with evolution, nor does it mean that a science teacher should present creationism or intelligent design as valid alternatives to the theory of evolution. It is perfectly appropriate for a science teacher to critique arguments for creationism or intelligent design that purport to be scientific (such as those listed two paragraphs earlier). However, in certain classes, depending on the comfort of the teacher in dealing with such issues and the make-up of the student body, it can be appropriate to address these issues. If questions about the validity of evolution or issues about creationism and intelligent design arise during science lessons, they can be used to illustrate a number of aspects of how science works and how scientific knowledge is built up over time, while always being open to the possibility of refutation and change.

My final suggestion is that science teachers do not get into theological arguments – for example, about the validity or interpretation of the scriptures; that is best left, in countries that have them, to lessons that are intended to deal with religion. However, many students enjoy hearing something of the personalities involved and there is much to be said for science teachers talking about the differing ideas and social circumstances of such luminaries as Mary Anning, Charles Darwin, Alfred Russel Wallace, T. H. Huxley, Samuel Wilberforce, Gregor Mendel and others, always acknowledging that few science teachers are specialist history teachers, so that there are dangers of stereotyping and oversimplification.

Cosmology

Less has been written about how science and religion might interact when teaching cosmology than when teaching evolution, but there is a developing literature. Bagdonas and Silva (2015) discuss a teaching approach which they developed and applied in a pre-service teacher education course. The course focused on the study of primary and secondary sources about cosmology and religion with primary sources written by cosmologists such as Georges Lemaître (the Belgian Catholic priest and scientist who argued that the universe had an abrupt beginning), Fred Hoyle (who argued in favour of the Steady State theory of the universe and invented the term 'the Big Bang', which he intended to be dismissive) and Pope Pius XII (who unsurprisingly favoured Lemaître's position). After studying these historical cases, the pre-service science teachers discussed how to deal with possible perceptions of conflict between scientific views and students' personal worldviews related to religion.

Many students enjoy discussing the anthropic argument. This argument exists in various forms, but the weak anthropic principle focuses on the fact that as physicists have learnt more and more about the universe, it is clear that the values of what are often termed the fundamental physical constants – things like the charge on the electron, the gravitational constant, Planck's constant and the speed of light – need to be remarkably close to their actual values if carbon-based life (and that's the only life we know of) is to have evolved. It is difficult to quantify 'remarkably', but a departure by less than just one part in a million, million, million, million would be more than enough to prohibit the evolution of even the most rudimentary carbon life forms (Briggs & Reiss, 2021).

What can we conclude from this 'fine-tuning'? Unsurprisingly, most physicists have not concluded that the universe is therefore designed by some divine being or set of beings, though this explanation cannot be disproved by science and is attractive to many with a religious faith. After all, it might be that there is only one universe, and we are fortunate in that it just happens to be one in which intelligent life could evolve. Another possibility is that the various fundamental physical constants are not independent of each other but connected in some way that physicists haven't (yet) established, so that the odds against intelligent life evolving are much smaller than the figure cited in the preceding paragraph. A further possibility is to take the anthropic principle at face value and conclude that all this talk of how remarkable it is that intelligent life has evolved is to have failed to take seriously the point that intelligent life is here and without it we wouldn't be having this discussion, so it's a logical fallacy to try to calculate the probability of life evolving without taking this into account – and if we take it into account, the probability is 1 (i.e., certain). Another possibility (the multiverse idea) is that there are actually lots of what we call 'universes', in the sense that communication between them is (probably) not possible, and that these universes may differ in their fundamental physical constants. In this case, there might be huge numbers of universes, only a small number of which allow intelligent life to evolve.

The human life cycle

A final example where science and religion can intersect in the science classroom is on the topic of the human life cycle. Religion plays an important role in the values that many people hold about start-of-life and end-of-life matters.

It is difficult to generalise with any great validity, as religions have within themselves a diversity of views, but a common position in many religions (e.g., Buddhism, Protestant denominations with Christianity, Hinduism, Sikhism) is that human life begins at conception. Accordingly, abortions are generally not permitted or are allowed only in exceptional circumstances – such as when the mother's life is in danger. Judaism has a particularly wide range of positions on abortion (depending on whether adherents belong to the progressive/liberal traditions or the more orthodox ones). Islam generally permits abortions early in pregnancy, as it is held that the soul does not enter the fetus until 120 days post-conception. Roman Catholicism teaches that any form of contraception, let alone abortion, is impermissible, though 'natural' contraception (the rhythm method) is allowed. Most religions also regard any form of euthanasia as unacceptable, and most religions teach that there is some form of life after death, and death is not seen as the end of life.

Conclusion

It is possible to ignore religion when teaching school science; in countries where religion is taught elsewhere in the curriculum, this position has some merit, so long as the colleagues in other subjects doing such teaching are able to teach well about the relationship between science and religion and to handle adequately scientific questions from students. In most countries, though, there is much of value in science teachers addressing issues to do with religion, as argued here, so long as the intention and anticipated outcome is that such teaching will aid students' education, especially their science education.

For the foreseeable future, many classrooms are going to have students who differ substantially in the importance they accord to religion. Teaching about religion in school science lessons therefore has risks. It can polarise and antagonise students, so that the last state is worse than the first. However, done well, it can help students better understand the nature of science and certain science topics, and even increase their understanding of those who have views about religion that differ from their own. It is difficult to imagine in today's world that this would not be a good thing.

Questions for further debate

1   How might science teachers avoid stereotyping and oversimplification when teaching about the history of science?
2   How might religion be taught about in school science in such a way that students better understand the nature of science and certain science topics?

Suggested further reading

Harrison, P. (2017). *The territories of science and religion.* University of Chicago Press.
Poole, M. (2007). *Exploring science and belief.* Hendrickson.

References

Bagdonas, A., & Silva, C. C. (2015). Enhancing teachers' awareness about relations between science and religion. *Science & Education*, *24*, 1173–1199. https://doi.org/10.1007/s11191-015-9781-7

Barnes, L. P. (2020). *Crisis, controversy and the future of religious education*. Routledge.

Behe, M. J. (1996). *Darwin's black box: The biochemical challenge to evolution*. Free Press.

Billingsley, B., Chappell, K., & Reiss, M. J. (Eds.). (2019). *Science and religion in education*. Springer.

Briggs, A., & Reiss, M. J. (2021). *Human flourishing: Scientific insight and spiritual wisdom in uncertain times*. Oxford University Press.

Cobern, W. W. (1996). Worldview theory and conceptual change in science education. *Science Education*, *80*(5), 579–610.

Commission on Religious Education. (2018). *Religion and worldviews: The way forward. A national plan for RE*. Commission on Religious Education. www.commissiononre.org.uk/wp-content/uploads/2018/09/Final-Report-of-the-Commission-on-RE.pdf

Coyne, J. A. (2009). *Why evolution is true*. Oxford University Press.

Kitcher, P. (1983). *Abusing science: The case against creationism*. Open University Press.

Kolstø, S. D. (2001). Scientific literacy for citizenship: Tools for dealing with the science dimension of controversial socioscientific issues. *Science Education*, *85*, 291–310. https://doi.org/10.1002/sce.1011

Laats, A. (2021). *Creationism USA: Bridging the impasse on teaching evolution*. Oxford University Press.

Levinson, R. (2006). Towards a theoretical framework for teaching controversial socio-scientific issues. *International Journal of Science Education*, *28*, 1201–1224. https://doi.org/10.1080/09500690600560753

Ling, T. (1980). *Karl Marx and religion in Europe and India*. Springer.

Mentzel, P. C. (2020). Introduction: Religion and nationalism? Or nationalism and religion? Some reflections on the relationship between religion and nationalism. *Genealogy*, *4*(4), 98. https://doi.org/10.3390/genealogy4040098

Meyer, S. C. (2010). *Signature in the cell: DNA and the evidence for intelligent design*. HarperOne.

Miller, J. D., Scott, E. C., & Okamoto, S. (2006). Public acceptance of evolution. *Science*, *313*, 765–766. https://doi.org/10.1126/science.1126746

Moore, R. (2007). The history of the creationism/evolution controversy and likely future developments. In L. Jones & M. Reiss (Eds.), *Teaching about scientific origins: Taking account of creationism* (pp. 11–29). Peter Lang.

Ofsted. (2019). *What is SMSC?* www.smscqualitymark.org.uk/what-is-smsc/

Oulton, C., Dillon, J., & Grace, M. (2004). Reconceptualizing the teaching of controversial issues. *International Journal of Science Education*, *26*(4), 411–423. https://doi.org/10.1080/0950069032000072746

Owens, D. C., Pear, R. S. A., Alexander, H. A. A., Reiss, M. J., & Tal, T. (2018). Scientific and religious perspectives on evolution in the curriculum: An approach based on pedagogy of difference. *Research in Science Education*, *48*(6), 1171–1186. https://doi.org/10.1007/s11165-018-9774-z

Pearce, P., Stones, A., Reiss, M. J., & Mujtaba, T. (2021). 'Science is purely about the truth so I don't think you could compare it to non-truth versus the truth.' Students' perceptions of religion and science, and the relationship(s) between them: Religious education and the need for epistemic literacy. *British Journal of Religious Education*, *43*(2), 174–189. https://doi.org/10.1080/01416200.2019.1635434

Pew Research Center. (2018). *How religious commitment varies by country among people of all ages*. www.pewforum.org/2018/06/13/how-religious-commitment-varies-by-country-among-people-of-all-ages/

Reiss, M. J. (1993). *Science education for a pluralist society*. Open University Press.

Reiss, M. J. (2008). Should science educators deal with the science/religion issue? *Studies in Science Education*, *44*, 157–186. https://doi.org/10.1080/03057260802264214

Reiss, M. J. (2018a). Why a chapter on religion in a book on science education? In L. A. Bryan & K. Tobin (Eds.), *13 Questions: Reframing education's conversation – Science* (pp. 295–311). Peter Lang.

Reiss, M. J. (2018b). Creationism and intelligent design. In P. Smeyers (Ed.), *International handbook of philosophy of education* (pp. 1247–1259). Springer. https://doi.org/10.1007/978-3-319-72761-5_86

Reiss, M. J. (2019). Evolution education: Treating evolution as a sensitive rather than a controversial issue. *Ethics and Education*, *14*(3), 351–366. https://doi.org/10.1080/17449642.2019.1617391

Sadler, T. D. (2009). Socioscientific issues in science education: Labels, reasoning, and transfer. *Cultural Studies of Science Education*, *4*, 697–703. https://doi.org/10.1007/s11422-008-9133-x

Stones, A., Pearce, J., Reiss, M. J., & Mujtaba, T. (2020). Students' perceptions of religion and science, and how they relate: The effects of a classroom intervention. *Religious Education*, *115*(3), 349–363. https://doi.org/10.1080/00344087.2020.1769537

Zeidler, D. (Ed.). (2003). *The role of moral reasoning on socioscientific issues and discourse in science education*. Springer.

Chapter 14

Sensitive issues in science

The case of relationships and sex education

Jenny Byrne

Introduction

> Relationships and sex education (RSE) has a reputation as a sensitive subject because of the intimate and personal aspects that are involved.
>
> (UNESCO, 2015)

In this chapter, I focus on relationships and sex education (RSE) as a sensitive issue. This is a controversial and challenging topic for many science teachers, and this chapter debates whether or not science teachers have a key role in dealing with and managing RSE. RSE is employed in this chapter to encompass Sex Education, Sexuality Education, Comprehensive Sex Education and Holistic Sex Education and other related terms that advocate the notion of establishing and managing respectful relationships. This central tenet of holistic and comprehensive approaches to RSE is recognised globally as good practice (UNESCO, 2018).

Science and sensitive issues

'Now, what I want is Facts. Teach these boys and girls nothing but Facts' (Dickens, 1854). Arguably, Mr Gradgrind's exhortation at the beginning of *Hard Times* is how science has traditionally been taught in school, where students are expected to learn 'the correct answer' to scientific problems (Levinson, 2006). Science has been regarded as an uncontroversial subject devoid of uncertainties and ambiguities, and this is reflected in science teachers' epistemic views and pedagogical practices (Levinson, 2006). But the science curriculum is replete with topics that are contentious and complex and do not necessarily provide a clear unambiguous answer (Owens et al., 2017). These topics can be broadly conceptualised as socioscientific issues; they are typically controversial and have moral, ethical, social, personal and political dimensions (Aivelo & Uitto, 2019; Reiss, 2019).

Whilst there is some overlap between controversial and sensitive issues in this chapter, sensitive issues are taken to be those that have an affective impact on an individual and connect to their emotions and personal feelings (Aivelo & Uitto, 2019; Reiss, 2019). Aivelo and Uitto (2019) comment that research into

DOI: 10.4324/9781003137894-17

teaching about sensitive issues has focused on health and physical education, rather than science. Lack of guidance can leave science teachers feeling vulnerable and ill-equipped about dealing with sensitive issues effectively (Oulton et al., 2004). However, science teachers have a key role in dealing with and managing health-related sensitive issues, such as RSE.

Science and health-related sensitive issues

Even though current changes to curricula, e.g., in England, Australia and New Zealand, have widened the disconnection between health and science, there is a consistent and broad consensus that personal health issues should be featured more prominently in science education, so that students can make scientifically informed decisions about their health behaviours (Roth, 2014; Zeyer & Dillon, 2014). Some (e.g., Arnold, 2018. Byrne & Grace, 2018; Harrison, 2005) argue that there is considerable overlap between the content of science and health curricula, making science a 'natural' home for teaching health-related topics.

The aims and goals of science and health education are complementary in addressing students' understanding of the real world, whilst also enhancing their scientific and health literacy. As Grace and Bay (2011, p. 2) state, 'Health literacy cannot be isolated from science literacy'. Zeyer (2012) concurs and considers that linking scientific knowledge with broader social issues is educationally advantageous in enabling students to become science and health literate.

Science and health are also closely related to students' everyday life. Using health-related issues in teaching some aspects of the science curriculum, particularly biology, provides a student-friendly approach, stimulating interest and curiosity in topics that have personal relevance and a real-world context (Byrne & Grace, 2018). In addition, an advantage in teaching sex education is that students' interest is usually high (Reiss, 2018). And yet, despite these advantages and the apparent synergies between science and health-related topics, it seems puzzling that sensitive health-related issues are not fully exploited in science lessons (Lee, 2012). The reasons for this are myriad, and an important point to note is the fear that teachers may feel about opening Pandora's Box when sensitive issues are introduced in science lessons. As one English trainee teacher commented, 'It's a minefield!' (Pickett et al., 2017, p. 334).

Why is school-based relationships and sex education necessary?

Parents are a child's first teacher and are ideally placed to offer advice about relationships and sex, but some are unwilling to initiate conversations about sex, or when they do the frame of reference can be inadequate or inappropriate (Goldman, 2008). Additionally, students and parents have been found to favour school-based RSE and think it is a necessary part of the curriculum (Depauli & Plaute, 2018; Peter et al., 2015; Saarreharju et al., 2020).

Schoolteachers are also ideally placed to deliver RSE because they know their students and the learning context; consequently, schools provide a secure and trustworthy environment where RSE can be taught systematically and comprehensively (Goldman, 2008; UNESCO, 2018). Furthermore, students' right to RSE is widely acknowledged. International research has recognised the benefits of high-quality, age-appropriate school-based RSE that specifically aims to improve adolescent sexual health and educational outcomes whilst facilitating young people's development as responsible, healthy and productive citizens (Goldman, 2008; Newby & Mathieu-Chartier, 2018; UNESCO, 2018).

Conversely, limited or complete absence of RSE in the curriculum leads to student ignorance that can have negative effects, such as lack of understanding and poor decision-making, higher rates of sexually transmitted infections, unwanted pregnancies and increased susceptibility to sexual abuse (Goldman 2008; Johnson et al., 2014; Stanger-Hall & Hall, 2011). Without school-based sex education, adolescents are likely to turn to alternative sources of information that may be less reliable or inaccurate (Reiss, 2018). It seems that RSE is a vital component of school curricula, but what is it and what are its aims?

What is school-based relationships and sex education?

Globally, RSE differs in terms of policy, content, the status of the subject, the age students start learning about RSE, the terms used to identify programmes and who is responsible for teaching it. However, a holistic definition of RSE is provided by UNESCO that states, regardless of national difference, the essence of Comprehensive Sex Education (CSE) is:

> a curriculum-based process of teaching and learning about the cognitive, emotional, physical and social aspects of sexuality. It aims to equip children and young people with knowledge, skills, attitudes and values that will empower them to: realize their health, well-being and dignity; develop respectful social and sexual relationships; consider how their choices affect their own well-being and that of others; and understand and ensure the protection of their rights throughout their lives.
>
> (UNESCO, 2018, p. 12)

Most Western countries now acknowledge that RSE should include more than the functionalist approach to sex education that previously focused on the 'plumbing' aspects of reproduction and disease prevention, and there has been a consistent trend to include a broader scope of information in RSE curricula (Peter et al., 2015; Garland-Levett, 2017). However, theory and policy are not always translated into practice. Janssens et al.'s (2020) systematic review found the content of most programmes is still focused on sexual risk prevention and has a heterosexual bias, with information about LGBTQI+ issues largely being excluded. Pedagogical approaches were found to be didactic and teacher-led with little evidence of addressing students' needs. RSE is frequently cited by students as 'too little too late', or as Selwyn and Powell (2007, p. 230) state, 'too little, for not long enough'. This disconnection or 'gap' between official sex education and students' interest and experiences of sexuality has been well documented (McKee et al., 2014). For example, students in Chicago (Jarpe-Ratner, 2020) felt that that the current content of RSE was far from adequate, particularly with regard to LGBTQI+ students. The focus on procreative sex was limiting, while the complete lack of discussion of pleasure was problematic. McKee et al., (2014) found that young people in Australia used the term 'scientific' to describe school RSE and regarded it as synonymous with 'irrelevant'. It conveyed no information about sexual pleasure that Allen (2005) describes as erotic forms of knowledge. Students were given the impression that sex is a bad thing and should be avoided, even though programmes teaching abstinence have been shown to be ineffective or even counterproductive (Stanger-Hall & Hall, 2011). But being prepared for safe sex, e.g., buying condoms, was frowned upon, leaving young people more likely to engage in risky sexual practices. A more holistic approach to RSE that includes information about sexuality as well as biological information would be less likely to be dismissed as 'just . . . the scientific side of things' (McKee et al., 2014. p. 661).

In contrast, good practice in RSE accords with liberal and democratic educational values that aim to develop students as learners for life, rather than treating them as repositories of (biological) knowledge, and to encourage a broader set of ideas, values and attitudes to flourish (Dewey, 1916; Reiss, 2018). This emancipatory approach includes listening to what students want included in RSE, so that a meaningful and holistic programme can be developed that empowers students to make informed healthy choices about their sexual lives (Allen, 2005; Garland-Levett, 2017). This will involve explorations of sex and sexuality, gender and gender norms, LBGTQI+ issues, pleasure, discrimination including homophobic bullying, sexting and sexual violence (UNESCO, 2018). It is therefore unsurprising that these sensitive topics may potentially lead to objections by parents and other stakeholders to teaching RSE in schools.

Objections to teaching RSE in schools

Opposition to RSE curricula that have a more liberal content has been voiced worldwide by those with particular cultural values, religious beliefs or political agendas. Goldman (2008) notes that the concerns of parents about RSE are often emotive and fear-based, rather than rational objections. In the Canadian province of Ontario (Saarreharju et al., 2020, p. 609), when the new sex education curriculum was introduced, it 'aroused downright froth' among the citizens. Such was the fervour of public demonstrations from parents who wanted to retain the sole right to teach their children about sex, and so fierce was the opposition from religious groups claiming that it did not correspond to their values, that it became a thoroughly politicised issue. Disapproval of RSE curricula is not limited to Canada. In England, parental protests occurred when RSE became mandatory with particular religious interest groups challenging schools' autonomy to decide how to manage relationships education for primary pupils (Vincent, 2020). Whilst in the USA, Peter et al. (2015) note concerns about teaching RSE due to fears of a backlash from a vocal minority of parents, and Hirose (2013) found that the public in Japan considered sex education to be too liberal. It is clear that RSE is a contentious issue, and that the complex and contradictory norms held by different cultural and religious groups regarding expressions of sexuality make decisions regarding what to teach in RSE challenging, which can create obstacles to teaching RSE (Janssens et al., 2020).

Obstacles to teaching RSE

Even though schools have long been identified as an appropriate place for RSE, there are many obstacles to securing high-quality RSE in schools, not least the fears and anxieties of teachers themselves and their concerns about violating taboos (Pound et al., 2017). In many countries, restraints on RSE may also be due to systemic or constitutional political factors, such as conservative school and state policies, or the ethos of the school pertaining to religious and cultural values that endorse restrictive school practices (Byrne et al., 2018; Gill, 2015). One fundamental aspect affecting the provision of RSE is that it is not necessarily compulsory in schools and in many countries, it has a low status. This means that RSE may not be prioritised by school management due to the competing demands of the curriculum and assessment regimes, or it may be that teachers regard RSE as irrelevant, particularly in relation to their professional development and promotion potential (Martínez et al., 2012, Pickett et al., 2017; Ram & Mohammadenzhad, 2019; Xiong et al., 2020). RSE is also frequently treated as a cross-curricular subject that can become untraceable or lost within the curriculum, with no one seeing it as part of their role or responsibility (Byrne et al., 2015; Martínez et al., 2012).

When faced with having to teach RSE, many teachers are reticent and would prefer not to engage, and this remains a challenge for effective RSE worldwide (e.g., Depauli & Plaute, 2018; Newby & Mathieu-Chartier, 2018; Martínez et al., 2012; Oerton & Bowen, 2014; Johnson et al., 2014; Ram & Mohammadenzhad, 2019). The barriers, perceived or real, to teaching RSE are manifold and ubiquitous; they include lack of training resulting in lack of knowledge and confidence and embarrassment. The pedagogical challenges of dealing with students' questions, insufficient teaching time, scarcity of resources, limitations imposed by the curriculum and assessment and perceived parental objections all compound the difficulty.

Firstly, teachers feel unqualified and inadequately prepared to teach RSE due to a lack of pre-service and in-service training (Byrne et al., 2015; Pound et al., 2017). Pre-service teacher training courses do not prioritise training in health, and particularly sensitive issues such as RSE and continuing professional development opportunities are rare or non-existent (Byrne et al., 2018; Oerton & Bowen, 2014). The lack of knowledge, expertise or confidence to handle issues in RSE, which are by their nature associated with large amounts of often incomplete information with no 'correct' answers, can be disconcerting for science teachers who are more used to dealing with factual certainties.

Secondly, lack of training not only leaves gaps in teachers' knowledge about how to deal with RSE in the classroom, but also leads to a lack of confidence about what to say, how to handle questions from students and fear of the consequences of making mistakes (Goldman, 2011; Johnson et al., 2014). Fear of embarrassment about what to say and how to say it may also be a stumbling block even when the content of the programme mainly focuses on basic biology. These feelings are likely to be exacerbated when other, more sensitive aspects of RSE are included (Pound et al., 2017; Reiss, 2018).

Some teachers are uncertain about where boundaries lie and what they are 'allowed' to discuss in class, and lack of guidance about this can inhibit teachers' confidence to deliver RSE (Pickett et al., 2017; Johnson, et al., 2014). Teachers have concerns about 'stepping over the imaginary line of 'appropriate content' and 'getting it wrong'. They fear negative reactions from parents when this conflicts with parental views and attitudes; for example, accusing them of teaching immoral content to children when teaching about sexual orientations (Pickett et al., 2017; Johnson et al., 2014, p. 371; Ollis, 2010; Ram & Mohammadenzhad, 2019). Added to this are teachers' own beliefs, value systems and culture that impact their attitudes towards RSE, with some reluctant to include social issues as part of (science) teaching (Aivelo & Uitto, 2019; Byrne & Grace, 2018).

Insufficient teaching time in an already crowded primary and secondary science curriculum is also regarded as a significant barrier in delivering RSE (Johnson et al., 2014; Martínez et al., 2012). Additionally, the curriculum and national assessment systems often give priority to acquiring knowledge and understanding of concepts, rather than providing time for consideration of social and ethical issues (Garland-Levett, 2017). The culture of performativity and academic achievement that prevails in many schools worldwide inevitably affects how teachers view their role, and may mean that teachers focus on factual biological information because of pressure to meet the demands of national testing, rather than holistic RSE that is not formally examined and frequently not assessed (Keogh et al., 2018; Selwyn & Powell, 2007).

Scarcity of resources is commonly cited as a reason not to teach RSE. When teachers feel uncomfortable about teaching sensitive issues, they resort to prepared materials, such as worksheets or textbooks, rather than interactive methods, such as classroom discussion, or they leave out key sections of a programme (Evans & Evans, 2007; Newby & Mathieu-Chartier, 2018). Reliance on external providers who 'parachute in and out' sometimes only once a year to give a set presentation is another way teachers avoid having to deal with RSE, but is regarded as inadequate in providing comprehensive RSE (Goldman, 2011; Martínez et al., 2012).

Avoiding teaching RSE for these reasons does not support students' wellbeing, intellectual, scientific or personal development; instead, it does them a disservice. Equipping teachers with the appropriate knowledge and skills to teach RSE effectively and manage potentially sensitive issues is essential (UNESCO, 2015). Specialised training has been shown to be effective in improving confidence and competence to manage and deal with RSE and is important at pre- and in-service levels as part of ongoing professional development (Byrne et al., 2018; Johnson, et al., 2014; Ollis, 2010).

Supporting good-quality SRE through science

Good science teaching should challenge students and enable them to think critically by engaging in issues that are relevant to them, such as RSE, and the science curriculum is uniquely suited to engage with RSE in meaningful ways (Arnold, 2018; Gill, 2015; Harrison, 2005). However, RSE is a multifaceted topic, and dealing with the biological and anatomical concepts in isolation from wider social and personal contexts makes the science abstracted from real life, and risks disengaging students as learners and failing to meet their educational needs (Allen, 2005). Roth (2014) argues for approaching science education through personal health, because it makes the science more relevant to students' lives. That is not to say that science is expected to deal with every element of RSE. Having the support of senior management, a positive school ethos towards RSE and working in collaboration with other subject areas, e.g., RE, health education and citizenship, RSE programmes can be devised that are empowering for all students (UNESCO, 2018). Some suggestions, although not exhaustive, for how this may be developed in science are discussed later.

It is important for all students know and understand basic biology with respect to human sexual development and reproduction, so it is understandable that primary school science and secondary biology portray sexual identity as a straightforward binary of male/female, men (boys)/women (girls); feminine/masculine (Harrison & Ollis, 2015). However some students do not identify with the 'normal' sexual orientation of their sex, or the biological gender recorded at their birth, or even a specific gender. It is therefore equally valid to understand sex determination, sexual orientation and gender more broadly to include those students who do not fit neatly into these heteronormative categories and challenge the *status quo* in how these topics are taught. This will ensure an inclusive and non-judgmental ethos that more closely reflects 21st century society. It will also take greater account of all students' needs however they may identify sexually (Bragg, et al., 2018).

In the early years of primary school, RSE focuses on good and bad relationships and simple human anatomy, including genitalia. This is dealt with as male and female and encourages a straightforward understanding of sexual difference, but teaching need not be solely related to physical appearance. Counteracting gender stereotypes and allowing children to understand how they feel, regardless of their gender or sexual orientation, will provide a good foundation in RSE.

In later primary and early secondary school, a more sophisticated understanding can be developed when topics focus on puberty and sexual development. Puberty is often taught in science lessons as a biological phenomenon and a sign of reproductive maturation that does not take into account any understanding of non-heterosexual or 'non-reproductive' bodies (Harrison, 2005, p. 73). This biomedical model is far removed from students' personal feelings and social experiences of puberty. It is often a particularly challenging period emotionally and physically for all students, especially for adolescents who are intersex or questioning their gender identity or expression (UNESCO, 2018). Science is ideally placed to help students make the connections between the biological processes and their personal and social experiences of puberty (Harrison, 2005). This may include challenging the narrative that puberty is regarded positively for males with the onset of sexual desire, power and virility, whereas for females it can be one of shame and confusion, due to conflicting messages about sexuality, virginity, fertility and womanhood (UNESCO, 2018). Puberty should be considered an important and valuable rite of passage to adulthood, regardless of gender identity or sexual orientation. Science can do much to promote this attitude by including discussions that challenge gender norms of masculinity and femininity and body image during lessons about the biological changes at puberty. For example, menstruation in some cultures is a taboo and girls are stigmatised, and often forced to stay away from school. Science teachers can help to ease this prejudice by ensuring menstruation is seen as a normal, healthy process in the same way as the production of sperm. It is also represented in biology text books as a failure (not to get pregnant), rendering women/girls as inferior to men/boys, which exacerbates gender power relations and constrains how students experience and express their sexuality (Garland-Levett, 2017). Facilitating a better understanding of sexuality and improving students' health and scientific literacy could involve engaging in critical reflection on the perceived normative attitudes about gender and sexuality (Harrison & Ollis, 2015) This could include reflection on how males and females are represented in general school discourse, as well as critiquing information sources, such as science textbooks, that in many countries tend to omit holistic aspects of sexuality (García-Cabeza & Sánchez-Bello, 2013; Reiss, 2018).

In secondary school, sexuality could be explored as a continuum, rather than the reductionist view of discrete categories, when teaching about the role of sex hormones on development. A better understanding of the physiology of these hormones and the role they play in sexual development could include acknowledging the close similarity in chemical composition of oestrogen and testosterone, and introducing students to the fact that both males and females produce the 'opposite' gender's sex hormones (Reiss, 2018). This could help to challenge heteronormativity and improve understanding for all students about issues related to sexuality and intersexuality, whilst supporting those students who identify as transgender or intersex. Discussion can lead students to recognise and accept differences by gaining a better understanding of sexual identity as well as sexual preferences, which may help to reduce prejudice and homophobic bullying.

Similarly, reproduction is presented as equating to sex education in the science classroom with its associated hegemonic, heteronormative values and attitudes (Reiss, 2018). Reproduction is recognised as the only legitimate purpose for sexual activity; this is not inclusive, as it effectively silences those students who identify as LBGTQI+ and can leave many students wondering about their sexual feelings, behaviour and relationships, and what 'normal' means (Jarpe-Ratner, 2020, p. 294). Acknowledging that sexual activity can be for pleasure and not just for procreation means that dialogue is opened up about safe sex practices, which is not solely focused on contraception, but includes information about STIs and HIV (Gill, 2015). In the USA, an increasing number of young people are engaging in risk-taking sexual behaviours because they believe that neither oral nor anal intercourse are really sex and that only penetrative vaginal intercourse is *really* sex (Harrison, 2005). Teaching about the mechanics of the condom as a prophylactic not only as a contraceptive, but as a barrier to disease, students can begin to understand how STIs occur, and this may also help to dispel myths about sexual activity, as well as combat prejudice and homophobic behaviour (Gill, 2015). Opening up discussions on condom use can facilitate opportunities to explore such ideas as peer pressure, gender stereotypes, rights and obligations, as well as the interrelated issues of alcohol and substance misuse with respect to sexual behaviour.

In conclusion, science, and in particular biology lessons, have a key role to play in RSE, and much good can be done by adopting alternative approaches to the prevailing norms exemplified in textbooks, worksheets and didactic teaching approaches. Opportunities for discussion, reflection and evaluation of their understanding of the key scientific concepts will support students' competence and confidence in using scientific knowledge effectively when appraising their sexual health and will ensure RSE no longer seems irrelevant to them (Arnold, 2018; Gill, 2015; Harrison, 2005). Lessons need to include more than the mechanics of heterosexual reproduction, contraception and disease prevention by putting greater emphasis on the relationship aspects of RSE: that will make lessons more inclusive and respectful for all students. This is emancipating for students and will empower them to make healthy decisions that will improve their long-term sexual health and educational outcomes. In doing so, it will enable them to flourish.

Questions for further debate

1   To what extent (and how) might the effectiveness of relationships and sex education be evaluated?
2   What should happen when a teacher is uncomfortable with discussing relationships and sex education?
3   What support do you think is needed at pre- and in-service levels to enable teachers to teach RSE effectively?

Suggested further reading

Janssens, A., Blake, S., Allwood, M., Ewing, J., & Barlow, A. (2020). Exploring the content and delivery of relationship skills education programmes for adolescents: A systematic review. *Sex Education, 20*(5), 494–516. https://doi.org/10.1080/14681811.

Jenkinson, A., Whitehead, S., Emmerson, L., Wiggins, A., Worton, S., Ringrose, J., & Bragg, S. (2020). *Good practice guide for teaching relationships and sex(uality) education. (RSE).* Institute of Education.

UNESCO. (2018). *International technical guidance on sexuality education.* The United Nations Educational, Scientific and Cultural Organization.

References

Aivelo, T., & Uitto, A. (2019). Teachers' choice of content and consideration of controversial and sensitive issues in teaching of secondary school genetics. *International Journal of Science Education, 41*(18), 2716–2735. https://doi.org/10.1080/09500693

Allen, L. (2005). *Sexual subjects: Young people, sexuality and education.* Palgrave Macmillan.

Arnold, J. C. (2018). An integrated model of decision-making in health contexts: The role of science education in health education. *International Journal of Science Education, 40*(5), 519–537. https://doi.org/10.1080/09500693

Bragg, S., Renold, E., Ringrose, J., & Jackson, J. (2018). 'More than boy, girl, male, female': exploring young people's views on gender diversity within and beyond school contexts. *Sex Education, 18*(4), 420–434. https://doi.org/10.1080/14681811

Byrne, J., & Grace, M. (2018). Health and disease. In K. Kampourakis & M. J. Reiss (Eds.), *Teaching biology in schools: Global research, Issues and Trends* (pp. 74–86). Routledge.

Byrne, J., Rietdijk, W., & Pickett, K. (2018). Teachers as health promoters: Factors that influence early career teachers to engage with health and wellbeing education. *Teaching and Teacher Education, 69*, 289–299. https://doi.org/10.1016/j.tate.2017.10.020

Byrne, J., Shepherd, J., Dewhirst, S., Pickett, K., Speller, V., Roderick, P., Grace, M., & Almond, P. (2015). Pre-service teacher training in health and well-being in England: The state of the nation. *European Journal of Teacher Education, 38*(2), 217–233. https://doi.org/10.1080/02619768

Depauli, C., & Plaute, W. (2018). Parents' and teachers' attitudes, objections and expectations towards sexuality education in primary schools in Austria. *Sex Education, 18*(5), 511–526. https://doi.org/10.1080/14681811

Dewey, J. (1916). *Democracy and education. An introduction to the philosophy of education.* Free Press.

Dickens, C. (1854). *Hard times.* Bradbury & Evans.

Evans, C., & Evans, B. (2007). More than just worksheets? A study of the confidence of newly qualified teachers of English in teaching personal, social and health education in secondary schools. *Pastoral Care in Education, 25*(4), 42–50. https://doi.org/10.1111/j.1468–0122.2007.00424.x

García-Cabeza, B., & Sánchez-Bello, A. (2013). Sex education representations in Spanish combined biology and geology textbooks. *International Journal of Science Education*, 35(10), 1725–1755. https://doi.org/10.1080/09500693

Garland-Levett, S. (2017). Exploring discursive barriers to sexual health and social justice in the New Zealand sexuality education curriculum. *Sex Education*, 17(2), 121–134. https://doi.org/10.1080/14681811

Gill, P. S. (2015). Science teachers' decision-making in Abstinence-Only-Until-Marriage (AOUM) classrooms: Taboo subjects and discourses of sex and sexuality in classroom settings. *Sex Education*, 15(6), 686–696. https://doi.org/10.1080/14681811

Goldman, J. D. G. (2008). Responding to parental objections to school sexuality education: A selection of 12 objections. *Sex Education*, 8(4), 415–438. https://doi.org/10.1080/14681810

Goldman, J. D. G. (2011). External providers' sexuality education teaching and pedagogies for primary school students in grade 1 to grade 7. *Sex Education*, 11(2), 155–174. https://doi.org/10.1080/14681811

Grace, M., & Bay, J. (2011). Developing a pedagogy to support science for health literacy. *Asia-Pacific Forum on Science Learning and Teaching*, 12, 2. www.ied.edu.hk/apfslt/

Harrison, J. K. (2005). Science education and health education: Locating the connections. *Studies in Science Education*, 41(1), 51–90. https://doi.org/10.1080/03057260508560214

Harrison, L., & Ollis, D. (2015). Stepping out of our comfort zones: Preservice teachers' responses to a critical analysis of gender/power relations in sexuality education. *Sex Education*, 15(3), 318–331. https://doi.org/10.1080/14681811

Hirose, H. (2013). Consequences of a recent campaign of criticism against school sex education in Japan. *Sex Education*, 13, 676–686. https://doi.org/10.1080/14681811

Janssens, A., Blake, S., Allwood, M., Ewing, J., & Barlow, A. (2020). Exploring the content and delivery of relationship skills education programmes for adolescents: A systematic review. *Sex Education*, 20(5), 494–516. https://doi.org/10.1080/14681811

Jarpe-Ratner, E. (2020). How can we make LGBTQ+-inclusive sex education programmes truly inclusive? A case study of Chicago Public Schools' policy and curriculum. *Sex Education*, 20(3), 283–299. https://doi.org/10.1080/14681811

Johnson, R. L., Sendall, M. C., & McCuaig, L. A. (2014). Primary schools and the delivery of relationships and sexuality education: The experience of Queensland teachers. *Sex Education*, 14(4), 359–374. https://doi.org/10.1080/14681811

Keogh, S. C., Stillman, M., Awusabo-Asare, K., Sidze, E., Monzon, A. S., Motta, A., & Leong, E. (2018). Challenges to implementing national comprehensive sexuality education curricula in low- and middle-income countries: case studies of Ghana, Kenya, Peru and Guatemala. *PLoS ONE*, 13(7), e0200513. https://doi.org/10.1371/journal.pone.0200513

Lee, Y. C. (2012). Socio-scientific issues in health contexts: Treading a rugged terrain. *International Journal of Science Education*, 34(3–4), 459–483. https://doi.org/10.1080/09500693

Levinson, R. (2006). Towards a theoretical framework for teaching controversial socio-scientific issues. *International Journal of Science Education*, 28(10), 1201–1224. https://doi.org/10.1080/09500690600560753

Martínez, J. L., Carcedo, R. J., Fuertes, A., Vicario-Molina, I., Fernández-Fuertes, A. A., & Orgaz, B. (2012). Sex education in Spain: Teachers' views of obstacles. *Sex Education*, *12*(4), 425–436. https://doi.org/10.1080/14681811

McKee, A., Watson, A.-F., & Dore, J. (2014). 'It's all scientific to me': Focus group insights into why young people do not apply safe-sex knowledge. *Sex Education*, *14*(6), 652–665. https://doi.org/10.1080/14681811

Newby, K. V., & Mathieu-Chartier. S. (2018). Spring fever: Process evaluation of a sex and relationships education programme for primary school pupils. *Sex Education*, *18*(1), 90–106. https://doi.org/10.1080/14681811

Oerton, S., & Bowen, B. (2014). Key issues in sex education: Reflecting on teaching, learning and assessment. *Sex Education*, *14*(6), 679–691. https://doi.org/10.1080/14681811

Ollis, D. (2010). I haven't changed bigots but . . . : reflections on the impact of teacher professional learning in sexuality education. *Sex Education*, *10*(2), 217–230. https://doi.org/10.1080/14681811003666523

Oulton, C., Dillon, J., & Grace, M. M. (2004). Reconceptualizing the teaching of controversial issues. *International Journal of Science Education*, *26*(4), 411–423. https://doi.org/10.1080/0950069032000072746

Owens, D. C., Sadler, T. D., & Zeidler, D. L. (2017). Controversial issues in the science classroom. *Phi Delta Kappan*, *99*(4), 45–49. https://doi.org/10.1177/0031721717745544

Peter, C. R., Tasker, T. B., & Horn, S. S. (2015). Parents' attitudes toward comprehensive and inclusive sexuality education beliefs about sexual health topics and forms of curricula. *Health Education*, *115*(1), 71–92. http://doi.org/10.1108/HE-01-2014-0003

Pickett, K., Rietdijk, W., Byrne, J., Shepherd, J., Roderick, P., & Grace, M. (2017). Teaching health education: A thematic analysis of early career teachers' experiences following pre-service health training. *Health Education*, *117*(3), 323–340. https://doi.org/10.1108

Pound, P., Denford, S., Shucksmith, J., Tanton, C., Johnson, A. M., Owen, J., Hutten, R., Mohan, L., Bonell, C., Abraham, C., & Campbell, R. (2017). What is best practice in sex and relationship education? A synthesis of evidence, including stakeholders' views. *BMJ Open*, *7*(5), e014791.

Ram, S., & Mohammadenzhad, M. (2019). Sexual and reproductive health in schools in Fiji: A qualitative study of teachers' perceptions. *Health Education*, *120*(1), 57–71. https://doi.org/10.1108/HE-02-2019-0005

Reiss, M. J. (2018). Reproduction and sex education. In K. Kampourakis & M. J. Reiss (Eds.), *Teaching biology in schools: Global research, issues and trends* (pp. 87–98). Routledge.

Reiss. M. J. (2019). Evolution education: Treating evolution as a sensitive rather than a controversial issue. *Ethics and Education*, *14*(3), 351–366. https://doi.org/10.1080/17449642

Roth, W.-M. (2014). Personal health – personalized science: A new driver for science education? *International Journal of Science Education*, *36*(9), 1434–1456. https://doi.org/10.1080/09500693

Saarreharju, M., Uusiautti, S., & Määttä, K. (2020). It goes beyond the fundamentals of sex and education. Analysis on the online commenting on the curriculum reform in Ontario. *International Journal of Adolescence and Youth, 25*(1), 609–623. https://doi.org/10.1080/02673843

Selwyn, N., & Powell, E. (2007). Sex and relationships education in schools: The views and experiences of young people. *Health Education, 107*(2), 219–231. https://doi.org/10.1108/09654280710731575

Stanger-Hall, K. F., & Hall, D. W. (2011). Abstinence-only education and teen pregnancy rates: Why we need comprehensive sex education in the U.S. *PLoS ONE, 6*(10), e24658. https://doi.org/10.1371/journal.pone.0024658

UNESCO. (2015). *Attitudinal survey report on the delivery of HIV and sexual reproductive health education*. The United Nations Educational, Scientific and Cultural Organization.

UNESCO. (2018). *International technical guidance on sexuality education*. The United Nations Educational, Scientific and Cultural Organization.

Vincent, C. (2020). The illiberalism of liberalism: schools and fundamental controversial values. *Journal of Ethnic and Migration Studies*, https://doi.org/10.1080/1369183X

Xiong, Z., Warwick, I., & Chalies, S. (2020). Understanding novice teachers' perspectives on China's sexuality education: A study based on the national pre-service teacher education programme. *Sex Education, 20*(3), 252–266. https://doi.org/10.1080/14681811

Zeyer, A. (2012). A win-win situation for health and science education: Seeing through the lens of a new framework model of health literacy. In A. Zeyer & R. Kyburz-Graber (Eds.), *Science|environment|health. Towards a renewed pedagogy for science education* (pp. 147–173). Springer.

Zeyer, A., & Dillon, J. (2014). Science|Environment|Health – Towards a reconceptualization of three critical and inter-linked areas of education. *International Journal of Science Education, 36*(9), 1409–1411. https://doi.org/10.1080/09500693

Chapter 15

Transition from primary to secondary

Sarah Earle

Introduction

A transition to a new stage of education can happen at many points in a child's life; for example, when moving to a new class or phase at the end of the school year. When a transition involves a move to a new school, it is also called a 'transfer'. This chapter will focus on the transfer from primary (or elementary) school to secondary (or high school) at around the age of 11 or 12, since this is a transition around which there is much discourse.

In this chapter, we will consider three areas of debate in turn:

- Is transition from primary to secondary school a 'rite of passage' or a threat to learning?
- Is transition the responsibility of primary or secondary schools?
- Is transition about a special project or developing a shared understanding?

Before reading further, it may be worthwhile to pause and consider your own responses to the three areas of debate. Are there any where you already have strong opinions? Do these arise from your own experience as a teacher, parent or child? By taking this moment to reflect, you can identify and make explicit your own preconceptions about transition.

Is transition from primary to secondary school a 'rite of passage' or a threat to learning?

The first question to consider is whether transition is really that much of a problem. For some, transition to secondary school could be seen as a 'rite of passage', a move to 'big' school, where a child develops a new level of independence as they move to the next stage of life. Such a move may involve travelling further and getting to grips with new teachers, new subjects and new routines. For a brief comparison of differences between primary and secondary school experiences, see Table 15.1.

DOI: 10.4324/9781003137894-18

Table 15.1 Typical changes from primary to secondary school in England

	Primary Ending with Year 6 (age 10–11), the final year of Key Stage 2 (age 7–11)	Secondary Beginning with Year 7 (age 11–12), the first year of Key Stage 3 (age 11–14)
Space	Often more local May involve drop off by parent/carer Smaller site Largely stay in one classroom	Often further away May involve travel on public transport Larger site Move between many rooms Specialist rooms
School staff and support	One class teacher who teaches most/all subjects Supervised at all times Smaller number of staff – likely to know school support staff, e.g., dinner supervisors and office staff	Different teacher for each subject Move independently between lessons Larger number of staff – less likely to know school support staff, e.g., science technicians and office staff
Peers	Consistent class/groups Smaller number of peers Live locally Oldest in the school	Consistent tutor group, but different groups/sets for different subjects Larger number of peers May live across a wide area Youngest in the school
Curriculum and pedagogy	More cross-curricular links between subject and topics National Curriculum for science split into year group topics, e.g., Animals including humans, Properties of materials, Light May be more child-focused pedagogy, e.g., start with the child	Subjects taught separately National Curriculum for science split into Biology, Physics and Chemistry May be more subject-focused pedagogy, e.g., start with the criteria

Changing schools could be a time for growth; transition can be seen as an opportunity for development (Gale & Parker, 2014). It could be argued that a move to 'big' school is a key part of growing up, in line with the changes brought with puberty. Moving to a larger school with specialist rooms and teachers is recognition that the child is becoming an adolescent, with a new level of independence and higher expectations of self-control. Rites of passage around this age feature in many religions, with the individual moving from childhood to preparation for adulthood.

However, whilst such changes may provide a growth opportunity for some, there is much evidence to suggest that transition to secondary school may result in learning regression for others. For example, Galton et al. (1999) estimated that

40% of pupils failed to make expected progress in the first year of secondary school. Evangelou et al. (2008) note that research consistently shows that attainment and attitudes can be held back or reversed. This regression is not something particular about the education system in England; Whitby et al. (2006) note this transition 'dip' is seen in other countries at the point of transition. A poor transition can also impact wellbeing and mental health (van Rens et al., 2018). This is compounded by an increasing attainment gap for pupils from low-income families, who make less progress than their more affluent peers (Social Mobility Commission, 2017).

For science in particular, negative attitudes are concerning, with low uptake of science in later years attributed to an 'engagement problem' (EEF, 2018). Archer et al. (ASPIRES, 2013) identified consistently low attitudes towards becoming a scientist from age 10, suggesting that children make their mind up about whether 'science is for them' at an early age. With the final year of primary in England often dominated by English and mathematics as the source of data for school accountability measures and school league tables, the low status of science may begin in primary. Wellcome (2017) found that across primary schools, the time spent on science was lower than the recommended two hours per week, whilst at the start of secondary school, in Key Stage 3 (age 11–14), Ofsted (2015) referred to as the 'wasted years', with repetition of work due to lack of communication with primary schools.

Many reasons have been explored to explain the long-standing issue of primary to secondary transition. Braund (2016) suggests that this negative impact on children's progress in science is due to factors such as repetition of work, lack of use of information about prior experiences and attainment and large changes in teaching environment and learning culture. He also notes a distrust between primary and secondary colleagues, which leads to a 'starting from scratch' policy in many schools. Sutherland et al. (2010) identify the lack of communication between primary and secondary schools as equivalent to 'two tribes', where there is a lack of trust between people who do not speak the same language. The 'different language' may even be prevalent in this chapter, with my background in primary schools identifiable in my use of 'child' and 'pupil' rather than 'student'.

In England, there is often no set 'feeder' secondary school, with children in an urban primary moving to up to 20 different secondaries, and a secondary school taking pupils from 60 different primaries. Galton and McLellan (2018) note increasingly distributed intakes due to academisation and changes in school admissions and parental choice. This increased variety of intake may explain why building relationships between primary and secondary schools is so difficult, and why pupils moving to secondary school arrive with such diverse experiences that it is seemingly impossible to value prior learning. It is therefore not a surprise that Ofsted (2015) found a lack of schools working together. Galton and McLellan (2018) examined transition across five decades and concluded that recent practice in England had regressed, resembling practice in the 1970s, due to an increasingly performative culture and the breaking up of local links with the move from Local Authorities to Academy Trusts.

A lack of communication between the 'two tribes' can result in the differences between primary and secondary education being so large that some children find it difficult to cope with the level of change. The Wellcome Trust (2009) suggested that the aim should be for a balance between continuity and discontinuity, where there is enough similarity to feel familiar, but there is also enough difference to signify a change with the move into adolescence. With much of the evidence falling on the 'threat to learning' side of the debate for the potential impact of a poor transition, the next question to consider is whose role it is to do something about it.

Is transition the responsibility of primary or secondary schools?

One area of debate for supporting science transition concerns the school personnel involved. If there is a problem with transition, then who should be called upon to act? In a primary school, transition could be the responsibility of the Year 6 (age 10–11) teacher or an Upper Key Stage 2 lead for a larger school, but this is unlikely to be the same as the person responsible for science in the school. In a primary school, the Science Subject Leader is a class teacher (for any year group) who supports science across the school by organising science curriculum resources and staff development. If the Science Lead teaches in Year 1 (age 5–6), then it will be logistically difficult to develop links with the secondary school. Likewise, in a secondary setting, the Head of Science is unlikely to be the person who is also responsible for organising transition activities, whilst the Year 7 (age 11–12) tutors and pastoral leads are unlikely to also have an interest in science. This lack of an identified person on each side of the 'divide' makes it difficult to run science transition activities and to build a closer relationship between schools.

The responsibility for a smooth transition should also require involvement from the child themselves or the parent/carer. However, van Rens et al. (2018) found that children and parents are not well represented in transition decision-making, but where there are positive relationships between schools, children and their parents, transfer challenges can be mitigated. With a large number of children on the move and limited resources, schools may decide to focus their energies on pupils who could be more likely to find the transition difficult. This could include children with identified special educational needs or disabilities (SEND), children from disadvantaged backgrounds, children who are working below the 'expected standard' or those with low levels of attendance. In these cases, the Special Educational Needs Co-ordinators (SENCo) and/or pastoral leads from each school may meet to discuss the child's needs, resulting in, for example, additional visits to the new school.

Beyond personnel, there are larger debates regarding the role of primary and secondary schools in science learning. Some may argue that 'real science' only starts at secondary school, hence taking a 'clean slate' approach to assume that little science teaching has taken place before arrival at secondary school (Braund,

2016). It could also be argued that science for younger children is primarily concerned with building positive attitudes, rather than developing extensive content knowledge. There are even those who propose that primary teachers are, on the whole, not science specialists, so should not teach scientific content which is likely to lead to misconceptions. However, children are exploring and inquiring about the world from the outset, building their own explanations for how the world works, and such 'common sense' frameworks may be at odds with current scientific thinking (Coppard, 2017). To leave such alternative frameworks unexplored and unchallenged until later years risks them being impossible to suppress (Bell et al., 2021).

The National Curriculum for England (DfE, 2013) contains a detailed list of objectives to be covered each year in primary science, and the reporting of science teacher assessment is statutory at the end of each Key Stage. Whilst time for science may have been squeezed by time spent on English and mathematics due to their use in school accountability measures, science does remain a 'core' subject in the primary school curriculum. There are increasing concerns that secondary school is too late to get children 'switched on' to science. In fact, there is much evidence to suggest that attitudes to science decline sharply in middle childhood, with pupils deciding early on that science is not 'for me' (Howard, 2017).

Children themselves may look forward to doing 'proper science' at secondary school (Davies & McMahon, 2011), with the stereotypical view of science as Bunsen burners and lab coats. This prevailing view may be strengthened by secondary school open evenings, where 'wow' experiments are promoted in order to recruit the visiting families to apply to join their school. This tactic may support enjoyment of open evenings or induction days, but such 'dramatic' experiments create unreasonably high expectations for secondary science lessons, with a resulting dip in science attitudes and motivation (Galton, 2002). Galton and McLellan (2018) found that the initial science lesson of naming the Bunsen burner looked similar in the 1970s and four decades later. Successful transition in science is the responsibility of both primary and secondary schools working together with families. All stakeholders need to be on board to develop strategies to support transition.

Is transition about a special project or developing a shared understanding?

Considering the amount of research and development in this field for at least the last five decades, why is transition still an issue? Researchers draw attention to the multifaceted nature of transition issues, meaning that a solution which only addresses one element will not provide lasting support. Many authors have framed the transition discussion in terms of five bridges (e.g., Galton et al., 1999; Howe, 2011; Galton and McLellan, 2018). The *administrative bridge* is focused on the transfer of information between schools, whilst the *pastoral bridge* considers the social and emotional needs of the child. The *curriculum bridge* considers

Table 15.2 Five bridges to support transition

Bridge	Examples
The administrative bridge	Arranging staff meetings and school visits to groups of pupils
	Provision of information to parents
	Transfer of pupil information between schools
The pastoral bridge	Meetings to consider social and emotional needs
	Open evenings or online tours
	Pupil and parent guides
	Induction days (where pupils visit secondary school)
The curriculum bridge	Curriculum mapping for progression and continuity
	Subject focused meetings or visits, e.g., lesson observation
	Joint activities or projects, e.g., 'bridging units' (discussed further later)
The pedagogical bridge	Cross-phase professional dialogue
	Meetings or visits focused on teaching strategies
	Developing common strategies for use across the transition
The student bridge	Supporting the development of self-management skills
	Extended induction programmes
	Including consideration of 'pupil voice' at all stages

continuity and progression, and the *pedagogical bridge* focuses on teaching strategies. The *student bridge*, perhaps the most underdeveloped in the literature, starts from the point of view of the child to consider their 'voice' in the transfer to secondary school. Table 15.2 provides some examples of these five bridges at work.

For transition in science education, the curriculum and pedagogical bridges are perhaps the most important, particularly with the debate mentioned earlier about the purpose of primary science and whether 'proper science' only starts in secondary school. Valuing what comes before as a foundation to build upon, rather than 'starting from scratch', could help to solve one of the key issues with transition, thus avoiding needless and boring repetition (Braund, 2016). Equally, for the primary teacher to be aware of what comes next, to instil the values of science and its place in the world is also key preparation for secondary school science. Thus, teachers on both sides of the 'divide' benefit from developing an understanding of each other's curriculum and pedagogical strategies.

What should be taught at each Key Stage is a much bigger debate beyond the scope of this chapter, but in England, the National Curriculum lays out the science content to be taught, so individual teachers can easily access the objectives for the Key Stage that precedes or follows theirs. However, a simple reading of the National Curriculum fails to capture the richness of the classroom experience, so it is likely that this step alone would do little to affect practice. Where individuals aim to build links with their local schools, or work together as school clusters or as part of an Academy Trust, there is much potential for cross-phase working. Those primary schools undertaking Primary Science Quality Mark (psqm.org. uk) are encouraged to build links with their local secondary schools. Secondary science teachers may actively link with their primary feeders through initiatives such as Ogden Trust Partnerships (ogdentrust.com) or the Great Science Share for Schools (greatscienceshare.org).

Making links between primary and secondary schools could involve:

- sharing of equipment or grounds;
- student visits, e.g., a STEM problem-solving day or a science share fair;
- meetings between staff, e.g., to discuss curriculum progression, sharing practice, moderation;
- staff visits to classrooms, to see science pedagogy in action;
- joint curriculum projects, e.g., 'bridging units';
- team-teaching of mixed-age groups.

To make the most of such links, especially those that 'cost' in terms of time and resources, the strategic purpose of the activities needs to be clear.

One way of developing stronger links between primary and secondary schools has been the use of 'bridging units', where activities are begun at the end of Year 6 (age 10–11) and continued into Year 7 (age 11–12). For example, one of the York STAY project (Braund, 2016) activities began with a letter to the children from a bakery requesting help with investigations into yeast and bread-making; these investigations were then continued and extended when the children started secondary school. Such bridging or transition units have been found to have an impact on the group working on them at the time (Braund, 2007), but despite the units themselves being made freely available online (pstt.org.uk), there is little evidence of their use beyond the development groups (Galton & McLellan, 2018). Davies and McMahon (2011) suggest that despite the apparent benefits of transition units, the practicalities of multiple feeder schools and the end of primary school focus on English and mathematics, together with pupil expectations about doing 'new things' at secondary school, mean that such units are rarely used. It is the collaboration of staff in the co-development of the projects that enhances the continuity between phases; this is lost when implementing a published bridging unit, so a pre-made transition unit has little influence on attitudes and practice (Davies & McMahon, 2011). It is possible for teachers and schools to develop their own bespoke units; for example, Qureshi created chemistry transition activities for her cluster related to the fiction book *Itch* by Simon Mayo (2012), which were used successfully that year (Qureshi & Petrucco, 2018). But other schools in the Trust were not able to continue the projects the following year, due to the amount of time and work involved. It appears that transition units can be used to build curriculum and pedagogy bridges, but that the benefits brought by such co-development can rarely overcome the practical and financial barriers.

Davies and McMahon (2011) describe a 'pedagogy gap', with primary school being more child-focused and secondary school being more concept-focused. Whilst teaching strategies may still differ, there would be many now who would question such a division between child and subject. The importance of eliciting pre-existing ideas, to both build upon and to identify misconceptions, are promoted as key for both primary and secondary (EEF, 2018). The growth in interest in cognitive science and educational neuroscience is provoking a change in thinking within some schools (e.g., Lovell, 2020). For example, a focus on knowledge development via strategies such as retrieval practice is becoming a central part of their practice, whether they are in primary or secondary education. There are those who suggest that this approach of direct instruction reduces the status of inquiry (Rogers, 2021), but this is a dismissal of 'discovery learning' rather than scientific inquiry, which is a key part of Working Scientifically in the National Curriculum. The role of practical work is a hot topic of debate in both primary and secondary school science, and whilst it is not the focus for this chapter, it is worth noting that utilising practical resources to help explore conceptual understanding is a key part of primary practice. For example, learning to name, describe and test the properties of materials like plastic, fabric and paper requires the experience of the physical materials. These resources could be used to develop scientific vocabulary through talk, or develop scientific methods through observation and measuring. This guided inquiry is not the same as 'discovery learning', so it is essential to clarify the meaning of such terms when entering into a debate about practical work in primary science. As with every debate, it is far more nuanced than cognitive science vs inquiry-based learning (see Chapters 8 and 9 for further discussion).

Work is ongoing to find ways to support transition from primary to secondary school. Some approaches have been captured by the Teacher Assessment in Primary Science (TAPS) project. The project has been working with teachers across the UK since 2013 to develop resources to support the teaching, learning and assessment of science in primary schools. One strand of the project considered the issue of primary to secondary transition. Pairs of primary and secondary schools took part in TAPS Transition (2016–19), although it was difficult to maintain consistent personnel and pairings due to the high turnover of staff (both changes in teacher roles and teachers moving to new roles in different schools). Some innovative practice was identified in the pairings, as exemplified in the case studies next.

Case study 1: building a pedagogical bridge – co-planning investigations

Transition events in the past had focused on other curriculum areas like sports, so Pauline Rodger from Holt Primary and Tom Daniels from St Laurence Secondary planned a summer term science event. Mixed groups of Year 6s and Year

Figure 15.1 Lolly stick catapults

7s worked together to investigate lolly stick catapults (Figure 15.1) in order to demonstrate their Working Scientifically skills. Students were given time to test prototypes then decide on their own success criteria for developing the catapult to perform 'better'. They then trialled the catapults, collecting data and drawing conclusions. Between groups, students peer-assessed the catapults and identified the possible next steps to improve against their criteria.

Pauline found that her Year 6 pupils were focused on the activity and kept setting themselves more challenges. Tom found that his Year 7s could work more independently than he had previously expected, giving him more time to spend with those who needed more support. Since then, Tom has rewritten his Key Stage 3 scheme of work to add 'buffer zones', which allow space for more self-guided tasks. Pauline spent further time comparing the Key Stage 2 and 3 curriculum, with a particular focus on use of vocabulary so that there is a more common language used across the schools.

Case study 2: building a curriculum bridge – using Focused Assessment

Sharon Heath from Trinity Primary and Maggie Beggs from Malmesbury Secondary worked together on an ENTHUSE project (with funding from STEM Learning), which allowed sustained interaction and visits. As part of their work,

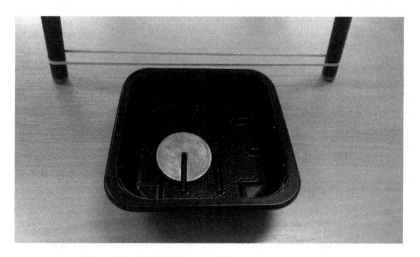

Figure 15.2 'Formula 1 margarine tubs'

they looked at progression in Working Scientifically and compared vocabulary and recording in science. As part of this, they explored the use of the TAPS Focused Assessment approach in Key Stages 2 and 3. This approach suggests the selection of one focus for attention and any pupil recording, within the context of a whole investigation, making formative assessment of practical inquiry more manageable. Sharon and Maggie developed a transition activity called 'Formula 1 margarine tubs' (Figure 15.2), whereby children investigate the effect of mass on distance moved. They found that mixed Year 6 and Year 7 groups recording in 'floorbooks' encouraged all students to contribute to planning and recording (with a different coloured pen for each, so that individual contributions could be identified).

Case study 3: building a student bridge – older children supporting younger children

Laura Smale from Abbeywood Secondary worked on science transition with her feeder primaries. She took a group of Year 7 students (from low-income backgrounds) to the primary school to support the younger ones with their data handling. The Year 5/6 pupils were investigating streamlining (TAPS aquadynamics activity) and had created their own tables to record results. The Year 7 students gave the Year 5 pupils feedback on their tables, for example, in their use of units and headings. The younger pupils then redrew their results tables under the guidance of the Year 7s. The Year 7s 'felt important' and developed their understanding of results tables in order to support the younger pupils. Laura has carried out similar activities with a 'blood splatter' activity, where pupils investigated the pattern between the height of the dropper and the diameter of the drop in order to identify a possible height for the 'attacker'. On this occasion, the Year 7s gave feedback on the graphs drawn by the younger children.

A key feature of all three case studies is collaboration between teachers and/or children. At the heart of TAPS is the notion that teaching, learning and assessment in science can be supported by developing a shared understanding of progression in scientific learning (TAPS, 2019). The TAPS Transition project has explored a range of ways to move from 'two tribes' to 'shared understanding', which will support a more coherent experience of science for learners. Key findings include the importance of dialogue between teachers to develop a shared language and active pupil involvement; for example, opportunities for Year 6 and Year 7 students to work together.

Conclusions

In this chapter, we have considered three debates in primary-secondary transition. In the first, it was accepted that some change is useful and necessary, but for many children, the transfer to secondary school can be a threat to learning, with progress stalling or regressing in the first year. This may be a particular issue for science, with high pupil expectations about secondary science experienced in the 'explosive' open evenings not being played out in the weekly Year 7 lessons, together with possible repetition of work due to lack of communication between phases. In the second debate, there was no single role or school upon whom the responsibility for science transition resides, which is perhaps part of the problem. It is in the interest of all stakeholders, including the family and pupils themselves, to be involved in discussion around smoothing the transition to secondary school.

The final debate, of what should be done to ease transition, is not so easily answered. Much research has gone into 'bridging units', but the practicalities of carrying out projects which continue across the 'divide' has meant that they are rarely completed, and 'off-the-shelf' solutions remove the collaboration element, which is key to the impact of this work. Other examples of transition work were explored in the TAPS case studies. These individual stories highlight the increasing need for a bespoke solution, which takes account of the individual context of the schools involved. For example, if a village school has one main feeder secondary, then it makes sense for the primary and secondary staff to work closely together, perhaps co-planning activities to involve Year 6 and 7 pupils. If it is a large urban secondary with myriad feeder primaries, then a shared understanding across schools is harder to accomplish, but online meetings between staff could help to develop awareness of curriculum progression.

Recognising that different strategies will work for different contexts is not the same as saying to start from scratch each time; there are still some key principles that can guide science transition decision-making:

- Aim to build a shared understanding between phases, if possible, spending time talking to colleagues across the 'divide' to break the 'two tribes' mentality – *the pedagogical and curriculum bridges*.
- Share information between schools, about children, curriculum and pedagogy, to aid progression rather than repetition – *the curriculum and pedagogical bridges*.
- Consider the balance of familiarity and change, if possible involving the children in such discussions – *the student bridge*.

- Plan for active student involvement in any transition activities; for example, Year 7/8 students supporting Year 5/6 pupils with investigations, or Year 6 and 7 students working together to solve STEM problems – *the student bridge*.
- Ensure communication between schools, pupils and parents – *the pastoral bridge*.

Consideration of areas of debate in primary to secondary transition has unsurprisingly revealed a complex picture, which is not easily solved. The key principles listed here provide a starting point for teachers and schools to examine their own practice and develop bespoke ways to smooth transition in science, to support both attainment and positive attitudes towards science which will recognise its key place in all of our lives.

Questions for further debate

1   What effective links can be made between primary and secondary school science?
2   How can these links be used to improve the transition process?

Suggested further reading

Braund, M. (2016). Improving continuity and progression from primary to secondary science. *School Science Review, 98*, 19–26.

Davies, D., & McMahon, K. (2011). Smoothing the trajectory: Primary-secondary transfer issues in science education. In A. Howe & V. Richards (Eds.), *Bridging the transition from primary to secondary school* (pp. 71–87). David Fulton.

Galton, M., & McLellan, R. (2018). A transition Odyssey: Pupils' experiences of transfer to secondary school across five decades. *Research Papers in Education, 33*(2), 255–277.

References

ASPIRES. (2013). *Aspires: Young people's science and career aspirations, age 10–14.* King's College London.

Bell, D., Mareschal, D., & The Unlocke Team. (2021). UnLocke-ing learning in maths and science: The role of cognitive inhibition in developing counter-intuitive concept. *Journal of Emergent Science, 20*, 19–26.

Braund, M. (2007). 'Bridging work' and its role in improving progression and continuity: An example from science education. *British Educational Research Journal, 33*(6), 905–926.

Braund, M. (2016). Improving continuity and progression from primary to secondary science. *School Science Review, 98*, 19–26.

Coppard, E. (2017). What can be learned about the teaching of the nature and behaviour of matter during the transition years from KS2 to KS3? Reflecting on the current science education research literature. *Journal of Emergent Science, 12*, 109–113.

Davies, D., & McMahon, K. (2011). Smoothing the trajectory: Primary-secondary transfer issues in science education. In A. Howe & V. Richards (Eds.), *Bridging the transition from primary to secondary school* (pp. 71–87). David Fulton.

Department for Education (DfE). (2013). *National curriculum in England: Science programmes of study*. Author.

Education Endowment Foundation (EEF). (2018). *Improving secondary science: Guidance report*. Author.

Evangelou, M., Taggart, B., Sylva, K., Melhuish, E., Sammons, P., & Siraj-Blatchford, I. (2008). *What makes a successful transition from primary to secondary school?* Effective Pre-School, Primary and Secondary Education 3–14 Project (EPPSE 3–14). Research Report DCSF-RR019. Department for Children, Schools and Families.

Gale, T., & Parker, S. (2014). Navigating change: A typology of student transition in higher education. *Studies in Higher Education, 39*(5), 734–753.

Galton, M. (2002). Continuity and progression in science teaching at Key Stages 2 and 3. *Cambridge Journal of Education, 32*(2), 249–265.

Galton, M., Gray, J., & Rudduck, J. (1999). *The impact of school transitions and transfers on pupil progress and attainment*. Research Report RR131. DfEE Publications.

Galton, M., & McLellan, R. (2018). A transition odyssey: Pupils' experiences of transfer to secondary school across five decades. *Research Papers in Education, 33*(2), 255–277.

Howard, S. (2017). Exploring the use of inquiry based science pedagogies across primary secondary transition: How does the literature relate this to the declining uptake of science in secondary schools? *Journal of Emergent Science, 12*, 99–108.

Howe, A. (2011). Exploring the great divide. In A. Howe and V. Richards (Eds.), *Bridging the transition from primary to secondary school* (pp. 1–13). David Fulton.

Lovell, O. (2020). *Sweller's cognitive load theory in action*. John Catt Educational.

Mayo, S. (2012). *Itch*. Random House.

Ofsted (2015). *Key Stage 3: the wasted years?* Ofsted.

Qureshi, A., & Petrucco, J. (2018). *Supporting transition, Primary science*, Special issue on the Teacher Assessment in Primary Science (TAPS) project.

Rogers, B. (2021). Does inquiry-based learning work? *Primary Science, 166*, 27–28.

Social Mobility Commission. (2017). *Low income pupils' progress at secondary school*. Social Mobility and Child Poverty Commission.

Sutherland, R., Ching Yee, W., & McNess, E. (2010). *Supporting learning in the transition from primary to secondary schools*. Bristol University.

TAPS (2019). *TAPS Transition: Teacher assessment in primary science (TAPS) support for primary-secondary transition*. Primary Science Teaching Trust.

van Rens, M., Haelermans, C., Groot, W., & Maassen van den Brink, H. (2018). Facilitating a successful transition to secondary school: (How) does it work? A systematic literature review. *Adolescent Research Review, 3*, 43–56.

Wellcome Trust. (2009). *Perspectives on education, Issue 2: primary – secondary transfer in science*. Wellcome Trust.

Wellcome Trust. (2017). *'State of the nation' report of UK primary science education*. Author.

Whitby, K., & Lord, P. with O'Donnell, S., & Grayson, H. (2006). *Dips in performance and motivation: A purely English perception?* QCA.

Chapter 16

Creativity in school science

Justin Dillon and Lindsay Hetherington

Introduction

Few among us would object to the idea that creativity is a desirable trait in humans; to be described as creative is usually seen as being a good thing. However, trying to identify what creativity looks like or how it might be taught and assessed is problematic. In this chapter, we look at debates about what counts as creativity and about the extent to which it can be taught and assessed. Our position reflects that of Morgan (2019), who wrote: "Creativity isn't the sole dominion of artists and geniuses. It shouldn't be confined to certain subjects in the curriculum, creativity in science is different to creativity in drama but is valuable to both. Creativity is enhanced through subject knowledge".

What is creativity?

Per Morten Kind and Venessa Kind describe creativity as "an elusive concept that is inevitably used by different workers in different ways" (2007, p. 2). Indeed, any paper, chapter, report, seminar or conference that addresses creativity invariably begins by attempting to define the term. Kind and Kind complain that it can become "a self-justifying label which is so 'politically correct' that it avoids critical scrutiny". This seems to us to be a bit of an overreaction, but we are sympathetic to the overall thrust of their argument.

Historically, conceptualisations of creation and creativity have changed considerably. So, while ancient civilisations might have seen the act of creation as being something that one or more gods did – making something from nothing – later understandings acknowledge that humans can be creative. Creativity was usually associated with the arts – creating a painting on a blank piece of paper, whereas now it is widely acknowledged that it can be found in many domains, including science (for a discussion of this issue, see Banaji et al., 2010).

Creativity in the curriculum

Around the turn of the century, there was significant interest in the role of creativity in the UK education system. Indeed, a review of the English national

DOI: 10.4324/9781003137894-19

curriculum, carried out in 2000, identified it as a key aim. Subsequently, the Qualifications and Curriculum Authority (QCA) was asked to investigate how schools could promote creativity (QCA, 2004). The resulting guidance for teachers, headteachers, senior managers and governors adopted, as a starting point, the definition of creativity devised by the National Advisory Committee on Creative and Cultural Education (NACCCE), which had published a seminal report five years earlier. This definition identified four characteristics of creative processes:

> First, they always involve thinking or behaving *imaginatively*. Second, overall this imaginative activity is *purposeful*: that is, it is directed to achieving an objective. Third, these processes must generate something *original*. Fourth, the outcome must be of *value* in relation to the objective.
>
> (NACCCE, 1999, p. 30)

Schools were encouraged to debate these characteristics as a "starting point for agreeing what your school actually means by creativity" (p. 7). The danger of this approach, of course, is that different schools would have different views of what counts as creativity. NACCCE had been explicit in defining creativity as "Imaginative activity fashioned so as to produce outcomes that are both original and of value". Some further insight comes in this statement:

> Creativity carries with it the idea of action and purpose. It is, in a sense, applied imagination. The imaginative activity is fashioned, and often refashioned, in pursuit of an objective. To speak of somebody being creative is to suggest that they are actively engaged in making or producing something in a deliberate way.
>
> (p. 31)

QCA went on to identify several outcomes associated with creativity: improvements to pupils' self-esteem, motivation and achievement. In addition, they argued that creativity prepares pupils for life and that it enriches their lives. None of these claims are supported with references to the literature (as is often the case with curriculum documents).

There is little in the QCA guidance that is specific to science, but there is one interesting clarification:

> When pupils are writing a poem, choreographing a dance or producing a painting, their work can be unique if it expresses their ideas and feelings. But what about work in subjects like science, history and maths? While it would be wonderful for a pupil to be the first person to discover a new scientific principle, this is highly unlikely. Does this mean that pupils can't be creative in these subjects? . . . Not at all. Skilled teachers can help pupils tackle questions, solve problems and have ideas that are new to them. This makes pupils' ideas original, the result of genuinely creative behaviour.
>
> (pp. 7–8)

Originality seems to be the key here to "genuinely creative behaviour" in science. QCA explicitly identify three different categories of originality:

Individual
A person's work may be original in relation to their own previous work and output.
Relative
It may be original in relation to their peer group: to other young people of the same age, for example.
Historic
The work may be original in terms of anyone's previous output in a particular field: that is, it may be uniquely original.

(p. 32)

This categorisation is broadly in line with Boden's (2004) idea of historic (h-creativity) and personal (p-creativity). Predating Boden's ideas, Anna Craft (2001) identified "big C creativity" (BCC) and ordinary, everyday, "little c-creativity" (LCC) – what she termed "possibility thinking". Hadzigeorgiou et al. (2012) align these ideas with Kuhn's (1970) terminology of "normal science" and "revolutionary science", which seems rather plausible. However, the QCA conceptualisation of originality seems to be at odds with Kind and Kind, who argue that "any approach to scientific creativity in school science be 'authentic' in scientific research terms, and meaningful in the school context" (2007, p. 3).

The Kinds' view seems to deny almost any opportunity for students to demonstrate creativity in science lessons (see also Hadzigeorgiou et al., 2012). This is a view that agrees with that of Moravcsik (1981), who opined that "Being creative is likely to imply being unconventional, unconforming, 'ahead' of the surroundings. In any given field, only a relatively few people have the ability and capability to be creative, and hence they are different from the others" (p. 225). We do not agree, and we are going to explore how creativity can be taught.

Can creativity be taught?

The NACCCE report is clear that "[c]reativity can be 'taught'" (1999, p. 11). It then goes on to identify two pedagogical dimensions: "Teachers can be creative in their own teaching; they can also promote the creative abilities of their pupils" (p. 11). Later it spells out a distinction between *teaching creatively* and *teaching for creativity*. "We define creative teaching in two ways: first, teaching creatively, and second, teaching for creativity" (p. 102). However, Jeffrey and Craft (2004), while acknowledging the value of distinguishing between teaching creatively and teaching for creativity, add a note of caution:

In making the distinction there is a danger that a new dichotomy becomes institutionalized in educational discourse, similar to those in the past such as formal and informal teaching or instruction and discovery learning.

(p. 77)

Their point is that "teaching creatively" might become associated with a normative "effective teaching" approach, whereas "teaching for creativity" might be seen as being more about "learner empowerment" (see also Jeffrey and Craft, 2004). Jeffery and Craft go on to argue that "a more useful distinction for the study of creative pedagogies would be the relationship between teaching creatively and creative learning" (p. 77). All of this discussion leads naturally to a consideration of what teachers can do to promote creative learning.

At this point, it is worth noting that J. P. Guilford, the American psychologist, spent considerable time looking at the relationship between creativity and divergent thinking. Runco (2001) points out that "Guilford's (1950) presidential address to the American Psychological Association often is credited with initiating the empirical research on creativity" (p. 245). This is incorrect, as Runco points out, but Guilford's work on creativity and on how to measure it played a major role in the development of the field. Perhaps "creative thinking" is a more useful term than "creative learning" in discussing teaching creativity.

So, what can science teachers do?

The teacher's role in promoting creativity has been discussed for some time. Garrett (1987), for example, says:

> For science teachers, there is a need to recognize the complexity of problem-solving and the concepts of originality and creativity it subsumes. This, in turn, demands that we provide situations within science classes that genuinely aim at developing the various skills and abilities associated with these activities.
>
> (p. 134)

Twenty-five years later, in a paper entitled "Creativity in science education", Daud et al. (2012) wrote that: "Creativity is a process that can be developed and enhanced. Every person has the potential; therefore, the potential should be enhanced by giving individuals the opportunity and the chance to [do] activities that enhance creativity" (p. 468). NACCCE, drawing on evidence from a wide range of sources, conclude that:

> The roles of teachers are to recognise young people's creative capacities; and to provide the particular conditions in which they can be realised. Developing creativity involves, amongst other things, deepening young people's cultural knowledge and understanding. This is essential both in itself and to promote forms of education which are inclusive and sensitive to cultural diversity and change.
>
> (p. 11)

In order to recognise young people's creative capacities, teachers need to be aware of what they might look like and how they develop. QCA (2004) suggested five ways that teachers might spot "pupils thinking and behaving creatively in the classroom". Children might be:

- questioning and challenging;
- making connections and seeing relationships;
- envisaging what might be;
- exploring ideas, keeping options open;
- reflecting critically on ideas, actions and outcomes.

Some years earlier, Frank Williams had proposed a Cognitive Affective Interaction Model which has three dimensions. Dimension 1 refers to the curriculum of the school; Dimension 2 refers to the teaching strategies used to promote critical thinking; and Dimension 3 refers to the cognitive and affective behaviours demonstrated by students. Dimension 3 is subdivided into two groups of factors:

> *Cognitive-Intellective factors*
> Fluent thinking: a significant number of relevant responses;
> Flexible thinking: a variety in the directions and categories of thinking related to questions, responses, etc.;
> Original thinking: unusual or novel responses that indicate originality;
> Elaborative thinking: additional detail or specificity in ideas or responses.
> *Affective-Temperament factors*
> Risk-taking: a willingness to make guesses, express ideas or take chances in investigating an unknown that might expose the individual to failure or criticism;
> Complexity: depth or intricacy in elaboration;
> Curiosity: an inquisitiveness about ideas, in problems and situations;
> Imagination: the ability to visualize and symbolize the unknown.

Williams' model can be used to devise curriculum activities (see for example Hsiao et al., 2006).

Hadzigeorgiou et al. (2012) offer a set of activities that they say "have the potential to increase the possibilities for students' creativity to emerge" (p. 609):

> *Creative problem solving* (e.g., measuring the height of a building using a barometer or tennis ball, measuring the surface of an irregular shape using a mechanical balance, the fate of the earth after the total disappearance of the sun, calculating the density of a proton, of a black hole).
> *Problem solving in the STS context* (e.g., how technology might affect the environment in the future, how we can produce electrical power in the future, how we might approach the sudden invasion of bacteria from space).
> *Creative writing* (e.g., a day in the life of a proton, a day without gravity).
> *Creative science inquiry* (e.g., investigating possible factors that might have an effect on the illumination of a room, the construction of a flashlight

from simple materials, ways to produce electricity for the house in a case of emergency, ways to heat water in the absence of metal containers).

Creating analogies to understand phenomena and ideas (e.g., the phenomenon of resonance, the ideas of energy, nuclear fission and fusion, chemical bonding).

Challenging students to find connections among apparently unrelated facts and ideas (e.g., what would be a connection between Newton's laws, a nurse and a soccer player? Between light, electrons and a surgeon? Between a glass of wine, the age of the universe and the evolution of stars? Between the sinking of the *Titanic* and hydrogen bonding?).

Mystery solving (e.g., detective work in order to explain the disappearance of something, like a certain volume of liquid; to find something that is missing, like a beam of light; to find connection between seemingly unrelated ideas, as in the case between a thief, the police and the speed of light).

Approaching the teaching and learning of science through the arts (e.g., using photography and making a collage to present the results of a study of a topic such as the effect of modern technology on everyday life; using technologies to construct scientific models; using drawing to represent a phenomenon, such as photosynthesis).

All these activities would seem to offer some affordances in terms of creativity, but they would require trialling and testing before it could be said that they actually do promote greater creativity. We will return to the science|art|creativity nexus later in the chapter.

What's the evidence?

Much writing about creativity in science education is rhetorical rather than evidence-based. Barrow (2010), for example, suggests a four-question approach drawing on the work of Cothron et al. (2006), which, he argues, can be used during inquiry-based lessons. The four questions are:

1   [What are the] Available materials?
2   [Are there] Different forms of materials?
3   What will be modified in attempting to answer the question?
4   How will its impact be measured?

Barrow argues, without providing any evidence, that the use of the four-question strategy allows "students to be creative in their designing a way to solve a testable question and helps science teachers' address their frustrations about inquiry" (p. 4). This assertion is itself testable but, at the moment, no one seems to have done it.

Empirical studies of creativity in science education are limited in their number. Cheng (2011) describes a three-year project involving more than 30 secondary science teachers across 30 schools in Hong Kong. The teachers took part in around 20 hours of professional development learning and trialling teaching strategies. Teachers could choose their own learning objectives, teaching strategies, science topics, and the pace and frequency of implementation. A range of methods were used to evaluate what happened. The teachers reported using a range of strategies, including "open inquiry, problem-solving, creative writing, making metaphors and analogies, creating drama, rewriting songs, [and] inventing new products" (p. 72). Many pupils, when asked about their experiences, reported "a more active learning style. They perceived the lessons to have less rote learning, reciting, teacher lecturing, but more activities, more practices, more student participation, more chances to interact with classmates and more self-initiated" (p. 72). However, there was very little evidence to "support that students could describe or criticize the creative thinking strategies, or know how to regulate or transfer them. In interviews, only very few of the students mentioned explicitly their conception and understanding of creativity and creative thinking strategies" (p. 76). The students were positive about the impact of the project on their science knowledge: "Many students considered their better understanding of science knowledge as their major learning outcomes. Students related this gain with their chance in discovering, presenting, and applying the science content knowledge" (p. 75). Cheng suggests that the students' views of the beneficial impacts of the new teaching approach might reflect the Eastern context of the study.

Cheng's study points to a challenge facing researchers in the area generally, which is identifying changes in students' creativity. Aktamis and Ergin (2007) discuss an initiative aimed at promoting scientific creativity, attitudes towards science and achievement in science. One class of 20 12 to 13-year-old students in Turkey was taught scientific process skills over a 12-week period. The topic focus was force and motion. The researchers used the Scientific Creativity Scale (Hu & Adey, 2002), which had been translated and adapted to the Turkish context. The scale used six items, whereas the original scale had seven. A pre-/post-test design was used, and a control group (n = 20) was used to see if any changes relating to the approach could be identified. The intervention group showed a bigger gain (pre: 17.30; post: 21.85) compared with the control group (pre: 15.65; post: 14.30). The study begs several questions, however, such as does the difference in the pre-test scores suggest that the groups were not well matched? Secondly, what is the explanation for the decrease in the score of the control group? The study adds little to our understanding of how to teach creativity.

The Hu and Adey (2002) Scientific Creativity Scale has seven items, including: "Please write down as many as possible scientific uses as you can for a piece of glass. *For example, make a test tube*", and "Suppose there was no gravity, describe what the world would be like? *For example, human beings would be floating*". The test was originally trialled with 160 students from three year groups (ages 11–12, 12–13 and 14–15). While the authors are positive about the potential of the survey, they do point out some limitations and the need for it to be validated by other researchers.

Ong et al. (2020) report on a five-day creative drama activity held in Taiwan, which was undertaken by 55 science majors and 28 non-science majors from five high schools in Malaysia. The research attempted to examine the effects of the intervention on situational interest, career interest and science-related interest. The authors also investigated students' perceptions of creative drama, and they report that it "was found to have triggered the situational interest in science within both majors". Other results included:

> The career interest and science-related attitudes of science majors were found to have significant improvement; some students' perception toward science careers and science have changed after the activity. Some students commented that creative drama had developed their courage, social skills, teamwork, creativity, self-reflection, presentation skills, critical thinking, and problem-solving skills.
>
> (p. 1)

This study, which, as the authors note, is small-scale, does have some interesting things to say about the value of creative drama as a means of teaching and learning science concepts.

Overall, then, the evidence base is patchy at best. There is a need for more work to be done to identify effective pedagogic strategies, as well as to explore what "effective" might actually look like. But is more knowledge about teaching creativity something that teachers actually want or need?

Educators and scientists' views of the need for creativity in science education

In a large, multi-national study reported by Hetherington et al. (2020), 270 educators, including teachers, trainee teachers, educators from out of school settings and teacher educators, took part in an online survey designed to explore their perceptions of the relationship between science and creativity. While there was broad agreement that science is a creative endeavour, a small number disagreed "about the relationship between science and creativity in the context of school science". The authors note that "The role of scientific knowledge within creativity in science education was found to be contentious" (p. 19).

Science, the arts and creativity

How might teaching creativity in science be different from teaching creativity in the arts? Our recent experience in the EU-funded Ocean Connections project provides some insight into the relationship between the two areas of the curriculum. Under the aegis of the Erasmus+ programme, Danish, English and Spanish higher-education institutions worked with aquariums and schools to devise an innovative blend of creative pedagogies and augmented/virtual reality aimed at teaching ocean literacy to students aged 7–14. For example, aquarium divers provided 360-degree video clips which students could annotate and incorporate into the virtual reality resources.

After a class of English primary students had visited a local aquarium, their teacher planned a follow-up lesson around a question that one of them had asked. The approach that the teachers used tended to be interdisciplinary rather than fully transdisciplinary. So, for example, a creative arts approach – making models – was used to stimulate students to think "as if" they were marine organisms in order to consider the difference between humans and marine organisms. The pedagogic strategies were all guided by a set of educative principles gleaned from comprehensive reviews of the literature. Another feeling is that the primary teachers in all three countries were happier with (that is, more used to) the idea of creativity in the arts, but were open to learning about it in science. Perhaps the new technologies got in the way as much as they offered affordances?

Chappell et al. (2019) discuss the relationship between science, the arts and creativity. Drawing on their work on the EU-funded CREATIONS project (see Chappell et al., 2015), they conceptualise creativity as:

> Purposive and imaginative activity generating outcomes that are original and valuable in relation to the learner. This occurs through critical reasoning using the available evidence to generate ideas, explanations and strategies as an individual or community, whilst acknowledging the role of risk and emotions in interdisciplinary contexts.
>
> (pp. 297–298)

Beginning with the assumption that the relationship is dialogic and relational, they examine how dialogue and material/embodied activity are manifested within creative pedagogies. The authors note that "the breadth of entanglement we were thinking with shifted, including entanglement between science|art but also with and between pupils, teachers, researchers, materials, environments, emotions and concepts" (p. 310).

Other examples of creativity in a science/arts context can be found on the website of the Boundless Creativity campaign, which was created by UKRI's Arts and Humanities Research Council (AHRC) in response to the Covid-19 pandemic (see https://ahrc.ukri.org/innovation/boundless-creativity/). An excellent example is "Dinosaurs and robots", a mixed-reality experience that combines storytelling and cutting-edge technology (see www.factory42.uk/dinosaurs-and-robots). The project is a collaboration between Factory 42, Sky, the Almeida Theatre, the Science Museum Group and the Natural History Museum. Using augmented reality, families are encouraged to explore, build and play games that combine STEM skills and creativity using mobile technologies.

Mike Watts outlines the case for "creative trespass" in science education, drawing on Arthur Koestler's observation in "The Sleepwalkers" of the "wrongheadedness of setting up academic and social barriers" between the sciences and the humanities. In 1998, Watts was involved in an international project which invited teachers and science educators who engaged pupils in the writing of poems or who themselves were writers of poetry to contribute to a collection (published as Watts, 2000; see also Watts, 2001).

Overall, then, we recognise that although creativity as an idea in education (and elsewhere) is hard to pin down, we do believe that it can be taught and probably assessed. However, the jury is still out.

Questions for further debate

1   In what ways can science education develop students' creativity?
2   How might science teachers develop their ability to teach creatively and for creativity?

Suggested further reading

Jeffrey, B., & Craft, A. (2004). Teaching creatively and teaching for creativity: Distinctions and relationships. *Educational Studies*, *30*(1), 77–87. https://doi.org/10.1080/0305569032000159750

National Advisory Committee on Creative and Cultural Education (NACCCE). (1999). *All our futures: Creativity, culture and education*. Author. http://sirkenrobinson.com/pdf/allourfutures.pdf

References

Aktamis, H., & Ergin, Ö. (2007). Bilimsel süreç becerileri ile bilimsel yaratıcılık arasındaki ilişkinin belirlenmesi. *Hacettepe Üniversitesi Eğitim Fakültesi Dergisi*, *33*(33), 11–23.

Banaji, S., Burn, A., & Buckingham, D. (2010). *The rhetorics of creativity: A literature review*. Creativity, Culture and Education.

Barrow, L. H. (2010). Encouraging creativity with scientific inquiry. *Creative Education*, *1*(1), 1–6. https://doi:10.4236/ce.2010.11001

Boden, M. (2004). *The creative mind: Myths and mechanisms*. Routledge.

Chappell, K., Hetherington, L., Keene, H. R., Slade, C., & Cukurova, M. (2015). *CREATIONS project deliverable 2.1: The features of creative inquiry learning*. www.creations-project.eu

Chappell, K., Hetherington, L., Keene, H. R., Wren, H., Alexopoulos, A., Ben-Horin, O., Nikolopoulos, K., Robberstad, J., Sotiriou, S., & Bogner, F. X. (2019). Dialogue and materiality/embodiment in science|arts creative pedagogy: Their role and manifestation. *Thinking Skills and Creativity*, *31*, 296–322. https://doi.org/10.1016/j.tsc.2018.12.008

Cheng, M. Y. V. (2011). Infusing creativity into Eastern classroom: Evaluations from students' perspectives. *Thinking Skills and Creativity*, *6*, 67–87. https://doi.org/10.1016/j.tsc.2010.05.001

Cothron, J., Giese, R., & Rezba, R. (2006). *Science experiments and projects for students*. Kendall-Hunt Publishing Company.

Craft, A. (2001) Little c creativity. In A. Craft, B. Jeffrey, & M. Leibling (Eds.) *Creativity in education* (pp. 45–61). Continuum.

Daud, A. M., Omar, J., Turiman, P., & Osman, K. (2012). Creativity in science education. *Procedia-Social and Behavioral Sciences*, *59*, 467–474. https://doi.org/10.1016/j.sbspro.2012.09.302

Garrett, R. M. (1987). Issues in science education: Problem-solving, creativity and originality. *International Journal of Science Education*, *9*(2), 125–137. https://doi.org/10.1080/0950069870090201

Guilford, J. P. (1950). Creativity. *American Psychologist*, *5*(9), 444–454.

Hadzigeorgiou, Y., Fokialis, P., & Kabouropoulou, M. (2012). Thinking about creativity in science education. *Creative Education*, *3*(5), 603–611. https://dx.doi.org/10.4236/ce.2012.35089

Hetherington, L., Chappell, K., Ruck Keene, H., Wren, H., Cukurova, M., Hathaway, C., Sotrious, S., & Bogner, F. (2020). International educators' perspectives on the purpose of science education and the relationship between school science and creativity. *Research in Science & Technological Education*, *38*(1), 19–41. https://doi.org/10.1080/02635143.2019.1575803

Hsiao, H. S., Wong, K. H., Wang, M. J., Yu, K. C., Chang, K. E., & Sung, Y. T. (2006). Using cognitive affective interaction model to construct on-line game for creativity. In *International conference on technologies for e-learning and digital entertainment* (pp. 409–418). Springer.

Hu, W., & Adey, P. (2002). A scientific creativity test for secondary school students. *International Journal of Science Education*, *24*(4), 389–403.

Jeffrey, B., & Craft, A. (2004). Teaching creatively and teaching for creativity: Distinctions and relationships. *Educational Studies*, *30*(1), 77–87. https://doi.org/10.1080/0305569032000159750

Jeffrey, B., & Craft, A. (2004). Teaching creatively and teaching for creativity: Distinctions and relationships. *Educational Studies*, *30*(1), 77–87. https://doi.org/10.1080/0305569032000159750

Kind, P. M., & Kind, V. (2007). Creativity in science education: Perspectives and challenges for developing school science. *Studies in Science Education*, *43*(1), 1–37. https://doi.org/10.1080/03057260708560225

Kuhn, T. (1970). *The structure of scientific revolutions*. University of Chicago Press.

Moravcsik, M. J. (1981). Creativity in science education. *Science Education*, *65*(2), 221–227.

Morgan, N. (2019). *The role of creativity in education*. Arts Council England. www.artscouncil.org.uk/blog/role-creativity-education

National Advisory Committee on Creative and Cultural Education (NACCCE). (1999). *All our futures: Creativity, culture and education*. Author. http://sirkenrobinson.com/pdf/allourfutures.pdf

Ong, K. J., Chou, Y. C., Yang, D. Y., & Lin, C. C. (2020). Creative drama in science education: The effects on situational interest, career interest, and science-related attitudes of science majors and non-science majors. *EURASIA Journal of Mathematics, Science and Technology Education*, *16*(4), 1–18. https://doi.org/10.29333/ejmste/115296

Qualifications and Curriculum Authority (QCA). (2004). *Creativity: Find it, promote it – Promoting pupils' creative thinking and behaviour across the curriculum at Key Stages 1, 2 and 3—practical materials for schools*. QCA.

Runco, M. A. (2001). Introduction to the special issue: Commemorating Guilford's 1950 presidential address. *Creativity Research Journal, 13*(3–4), 245–245. https://doi.org/10.1207/S15326934CRJ1334_01

Watts, M. (2000). Family science: Generating early learning in science. *Early Child Development and Care, 160*(1), 143–154.

Watts, M. (2001). Science and poetry: passion v. prescription in school science? *International Journal of Science Education, 23*(2), 197–208.

Chapter 17

Science and mathematics

The mathematical demands of science

Victoria Wong

Introduction

Mathematics is central to science, and learning how to use mathematical ideas, to reason mathematically and to understand mathematical concepts is a key part of learning science. While there are debates about how much or which mathematics is necessary for a scientific career, there is little debate that at least some mathematics is necessary to do science. Osborne (2014) argues that students should engage in scientific practice to improve the quality of their learning and identifies eight practices of science which should be in evidence in science classrooms, two of which are mathematical. These are analysing and interpreting data and using mathematical and computational thinking. However, many authors have identified that students have difficulty in using mathematics in science, both at school level (for example Dodd & Bone, 1995; Porkess, 2013; Needham, 2016) and at university level (for example Watters & Watters, 2006; Koenig, 2011; Redish & Kuo, 2015). Osborne (2014) argues that many science teachers do not believe that it is their responsibility to educate students in the mathematics required to understand science. He further argues that mathematics is important in science, but all too often this importance is not made manifest in either school science or science education research. This lack of research means that arguments made about students' use of mathematics within science are often not backed up by robust evidence.

There are three interconnected debates that I will discuss in this chapter:

- the nature of mathematics within science and how that compares to pure mathematics;
- the challenge of these differences for teachers, including the notion of transfer and the problems of working across the disciplines;
- the amount of mathematics which should be present in the science curriculum.

I will explore each of these debates and consider the implications that they have for science teachers and science education research.

DOI: 10.4324/9781003137894-20

Mathematics in science and in pure mathematics

Numbers and the meaning of symbols

Students are taught to use mathematical equations within mathematics and within science and at first glance they appear to be similar, for example $a = b \times c$. However, Redish (2017) argues that in physics, equations are linked to physical systems and that this adds information about how they should be interpreted. Science tends to use what can be called quantity calculus or values rather than pure numbers. Values have a unit as well as a number and these units impact whether and how those numbers relate. Thus, symbols in equations in science stand for measurements and the units relate to the measurement process (Redish, 2017). This has mathematical impacts which constrain the manipulation of the terms, compared to the manipulation of pure numbers. In the simplest terms, it is not possible to add two numbers if they have different units; one cannot add length to mass.

Redish and Kuo (2015) argue that the differences in the way mathematics is used across physics and mathematics are significant enough to amount to a difference in dialect or even a different-but-related language. They argue that physicists make meaning with mathematics in a different way to mathematicians, partly due to a difference in purpose; physics (and indeed science more broadly) is concerned with representing meaning about physical systems, where mathematics is more often concerned with abstract relationships. The need for values with units arises from this focus on physical systems, the real world, the context. As a result:

> Physicists 'load' physical meaning onto both symbols and numbers in a way that mathematicians usually do not. . . . It allows physicists to work with complex mathematical quantities without introducing the fancy math.
>
> (Redish & Kuo, 2015, p. 565)

In other words, the symbols (and their related units) are 'used to convey information omitted from the mathematical structure of the equation' (ibid., p. 567) and 'ancillary physical knowledge – often implicit, tacit, or unstated' (ibid., p. 568) is used when applying mathematics to physical systems. Using physical knowledge when applying mathematics to physical systems might involve knowing the limits to the numbers which can be put into an equation. For example, in the ideal gas equation ($PV = nRT$), none of the values would be negative as pressure, volume, amount of substance, the ideal gas constant and temperature in Kelvin are always positive values. There is no mathematical reason why a negative number could not be introduced into the equation, but there are reasons based on physical knowledge and the values which are represented by the symbols. Redish and Kuo (2015) argue that using this physical knowledge helps to keep the mathematics more straightforward.

Graphs and graphing

By the time they finish primary school at age 11, most children can make a graph to present numbers they have collected. Alongside presenting information, the value of graphs is their relational representation and predictive power, and it is these which are most commonly the focus of teaching about graphs in secondary education.

In school mathematics, a graph is often thought of as a representation of an exact relationship, which is described by a mathematical equation. Students are expected to be able to translate between algebraic and graphical representations. For example:

Draw the graph of $y = 0.8x$ for values of x from 0 to 6 (axes are given)
(AQA, GCSE mathematics, paper 3, June 2018)
Or:

Here are sketches of two graphs

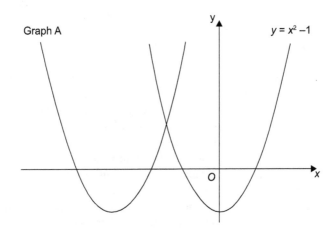

Graph A

$y = x^2 - 1$

The graph of $y = x^2 - 1$ is translated 3 units to the left to give graph A

The equation of graph A can be written in the form $y = x^2 + bx + c$

Work out the values of b and c.

Figure 17.1

(AQA, GCSE mathematics, paper 1, May 2018)

The emphasis in mathematics is on the graph as the representation of an exact or idealised relationship. It is emphasised that points which do not lie on the line do not fit the relationship in question, and there is therefore limited treatment

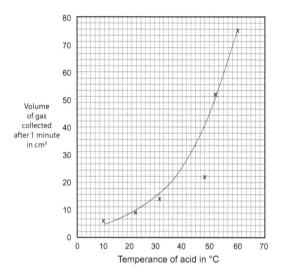

Figure 17.2[1]

of errors. In contrast, in science, students must deal with messy data that they or others have collected, and they learn about errors in data and natural variation, which may mean that not all points sit exactly on the line of best fit. For example, the graph above is given in a question following experimental details. Note that not all the points are exactly on the line.

Students are asked to draw a circle around the anomalous point, to suggest what may have happened to cause the anomalous result, and to explain how they could improve the accuracy of their gas collection.

The assumption is that all the points should lie on the line and that they do not is due to experimental error. Explaining where the error comes from and the effect on the data is part of learning to evaluate methods and explain uncertainty.

Thus, students may be asked to draw a graph of a function in a mathematics lesson, and in a science lesson to draw and explain a graph of their results. These are quite distinct tasks, and proficiency in one is by no means a guarantee of proficiency in the other. Furthermore, students who do not appreciate the distinction in what is asked of them in mathematics and in science may simply assume that the points which do not sit on the line in the graph are wrong, rather than the consequence of experimental error.

Leinhardt et al. (1990) review research into graphing in both mathematics and science and break down what students have to do into two types of actions: interpretation includes reading or gaining meaning from a graph; construction includes plotting a graph or determining an equation from a graph.

The different focus in mathematics and in science leads to students being asked to do very different graphing activities in each subject. In mathematics, interpretation might be looking at a given graph and determining where its x or y intercept is or calculating the gradient. Construction might involve sketching graphs of equations for a variety of functions including quadratics (for example, $y = \frac{1}{2}x^2 - 2$).

Two construction tasks are particularly relevant to science education: prediction and scaling. Prediction involves conjecturing from part of a graph where other points, which are not given, should be located. Leinhardt et al. (1990) note that not all prediction tasks require the same skills. Some are more reliant on estimation and perhaps measurement skills, whereas others depend on pattern recognition. Crucially for science, being able to predict requires that the student can interpret or understand the context, and what the representational graph should look like. For example, with a cooling curve, predicting the shape of the graph would require understanding that at the freezing point, the line will flatten until the entire sample has solidified and only then will the temperature of the sample fall further.

Scaling tasks are those where the focus is on axes, their scales and units. In virtually all graphs in science, the units are important because the graphs are representations of real-life contexts. The units are also often different on each axis, for example temperature against time, distance against time or light intensity against number of bubbles. Consequently, the scales will often be different on each axis, and learning how to choose and construct a sensible scale is a key part of learning to graph in science. As noted already, for the majority of the graphs drawn in mathematics there is little attention paid to what the variables actually represent, and thus there is limited focus on units. This leads to less attention being paid to learning how to choose and construct sensible scales.

Context

In both the use of equations and the use of graphs, the role of the situation or context in mathematics and in science is very different. In mathematics, a context, which may be a scientific context, can be taken from the real world to help students learn about graphs. The point of the questions asked is to support students in using mathematics, not the context itself. While mathematical ideas may be better understood in a context that students can relate to, the aim is for better understanding of the mathematics or mathematical function, rather than the context itself. The exception to this is distance-time graphs, where students are asked questions which are similar to those which might be asked in the physics curriculum; for example, to calculate speed from the gradient of the graph. Indeed, students are often taught by their mathematics teachers to ignore the context and focus on the mathematics, and there is discussion amongst mathematics educators as to whether including context actually helps the learner (Leinhardt et al., 1990).

This is very different from the situation in science, where the graph or an equation is viewed as an aid to understanding the phenomena being investigated through representing observations and aiding the detection of underlying

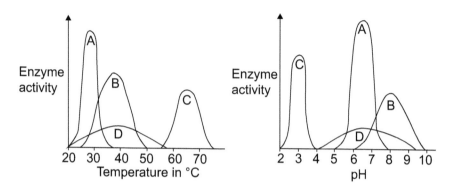

Figure 17.3 A – D are enzymes. Which is likely to have come from dark, hot, humid and acidic caves?

patterns. Consequently, interpretation of graphs may well involve shifting between the graphical representation to the situation or phenomenon. This is a less commonly required skill in mathematics education. For example, to interpret the graphs of enzyme action shown above requires an understanding of the effect of temperature and pH on enzyme action. Both an understanding of the graph and an understanding of the scientific situation are required to make sense of the data.

Transfer, collaboration and the difficulties for teachers

Transfer

Transfer is the use in one subject of what has been learnt in another. It is a contested notion in the research literature, with authors expressing a range of views as to what it is, whether it exists and if and how it can be promoted by education. Osborne (2014) argues that many science teachers believe that it is not their responsibility to educate students in the mathematics required to do science. Indeed, Goldsworthy et al. (1999) found that although science teachers recognised that students struggled with graphing, they did little to help them improve their skills in graph construction and interpretation. Likewise, in research carried out with Dutch teachers, Turşucu et al. (2017) found that the majority thought that the so-called transfer problem of students finding it difficult to use mathematics within science should be solved by more intensive practice in the mathematics classroom.

Boaler (2002) suggests that one's understanding of transfer is related to the theory of learning one subscribes to. She argues that both behaviourist and constructivist theories of learning, while very different, represent 'knowledge as a characteristic of people that may be developed and then used in different situations' (p. 42). A behaviourist view, for example, would lead to plenty of practice of mathematical methods, with the underlying assumption being that knowledge which has been clearly communicated and received would be available for future use, including in new contexts (Boaler, 2002). Schwartz, Bransford and Sears argue that constructivism by its very nature involves transfer, as 'whenever we assert that new learning builds on previous learning, we are assuming that some sort of transfer is involved' (2005, p. 11) and 'research on preconceptions involves paying attention to what people transfer in as it will profoundly affect what they learn' (p. 9). In contrast, Lave and Wenger (1991) argue that learning is situated, occurring in a context and a culture, and therefore students do not bring parcels of knowledge from mathematics with them into other contexts, including science lessons.

If, however, science educators hold the view that knowledge can be developed in one situation and then used in another, it would explain why they expect that the transfer of knowledge from mathematics to science will cause minimal difficulty for students. The assumption that transfer will be straightforward will tend to lead to deficit views of both students and mathematics teachers if, or rather when, the expected transfer is not seen.

Some authors likewise argue that transfer *should* happen and question why it does not, even suggesting students show an active resistance to transferring knowledge. Dodd and Bone mark 'the frequent inability or reluctance of pupils to transfer knowledge and understanding from the mathematics classroom to the science laboratory' (1995, p. 103). Lerman (1999) identifies the problem of transfer as what happens when people perform differently on what, to a mathematician, is the same task in different contexts. Larsen-Freeman suggests that it is a common problem that students 'fail to transfer their learning' where, although they appear to have learned something in one context, they cannot activate it in another (2013, p. 107). These authors all appear to hold a deficit view when comparing how students perform compared to how it might be hoped they would perform. Lobato (2012) suggests the deficit view holds sway as teachers do not notice when students *do* transfer their learning and, consequently, transfer is more often noted when it does not happen than when it does.

However, if it is the case that there are differences between mathematics and science in the types of task and purpose for both graphing and the use of equations, it follows that the teaching of these are necessarily different, with different foci. While there are aspects in common (for example, plotting points on a graph, understanding that a graph's axes represent different things and the significance of the gradient), the uses of graphs and equations in mathematics and in science are not the same. It is, therefore, not reasonable to expect students to be able to draw graphs in science using scientific conventions simply because they have also had to draw them in mathematics using mathematical conventions (or vice versa). Nor is it reasonable to expect students will be able to use and interpret scientific equations simply because they have also used equations in mathematics.

Another way of phrasing this is that teachers expect that students will 'know that' and 'know how' – and be able to use this knowing to solve problems without much additional teaching or assistance. This view of transfer is likely to lead to frustration, and a deficit view of both students and mathematics education. If instead students' prior mathematics study is seen as preparation for future learning (as argued for by Schwartz et al., 2005), then science teachers might expect that prior study of a topic in mathematics would make it easier for students to learn in science, although – and this is critical – they would still need additional teaching in that topic in science. An example of this might be that the prior study of proportional reasoning could make it easier to learn about moles, mole ratios and concentration, all of which depend on proportional reasoning. Crucially, teaching about the specific science and the application of proportional reasoning to that science would still be required.

Mathematics-science collaboration

I will use the term 'collaboration' here to describe any way in which mathematics and science departments work together, including formally and informally. There are many calls in the literature for collaboration (or integration or interdisciplinary work – the terms are frequently not clearly defined), including from Venville et al. (2002), and Honey et al. (2014). Reasons given for collaboration can be summarised into three main cateogories: that there is an overlap between the disciplines, such that collaboration is the logical response; the desire for improved transfer; and to meet an aim for an increase in student engagement.

However, there are numerous barriers to collaboration, including time, timetabling, physical distance between departments, the nature of the curriculum, the asymmetric dependency between mathematics and science, and blaming the mathematics department for students' difficulties with mathematics in science (Wong, 2018; Wong & Dillon, 2019).

What collaboration between mathematics and science departments can usefully achieve is for science teachers to understand what is (and, critically, what is not) part of the mathematics curriculum and therefore what they will need to teach within science. In other words, such collaboration can help to recalibrate science teachers' expectations of their students and support them in changing their practice to include explicit teaching of mathematics within science. This is an undoubted benefit for science teachers and the science department; the benefit for mathematics teachers in such a scenario is harder to ascertain. Hence, I use the term 'asymmetric dependency' to highlight the discrepancy in benefit for each department. Science is dependent on mathematics; mathematics is not dependent on science; and thus, science education will tend to be the main beneficiary from such collaboration. Unless this asymmetric benefit is explicitly acknowledged, this type of collaboration, however useful to science teachers, will tend to collapse (Wong, 2018; Wong & Dillon, 2019).

The challenge for science teachers

The challenge for science teachers remains that many students find using mathematics within science hard, particularly if the goal is for them to reason using mathematics. Research provides few useful or concrete suggestions as to how teachers might support the development of their students' mathematical thinking.

The national curriculum for science in England changed in 2015, and the external assessments which all students sit at the end of year 11 (age 16) must now include a government mandated minimum percentage of mathematics. The majority of students sit a combined science qualification (a GCSE), which includes aspects of biology, chemistry and physics. The minimum percentage of mathematics for this qualification is 20% (Ofqual, 2015).

As part of that curriculum, students must be able to recall more than 16 different equations in physics, plus some equations for chemistry and biology. They must be able to use a further seven physics equations which are given on an equation sheet.

There are two separate tiers which can lead to the award of a GCSE in science: higher tier and foundation tier. In practice, in foundation tier the equations are usually given to students as so few of them can recall as well as use them. As a result, students are frequently given an equation and some numbers and have to put the numbers into the equation to generate a numerical answer. For example:

There was a current of 0.020 A in the resistor for 180 seconds.
Calculate the charge flow through the resistor.
Use the equation: charge flow = current × time

Students are also given the units for charge flow. There are no follow-up questions to this; students are not required to 'think mathematically' or to reason using mathematics. Students simply substitute the numbers into the equation and plug them correctly into the calculator. No understanding of charge flow, current or time is required, and students are not expected to relate this calculation back to the context. As such, I would argue that it is questionable that this is actually science. It does not really fit in with Osborne's (2014) science practice: using mathematical and computational thinking. It is fulfilling a government mandate that 20% of a GCSE external assessment for 16-year-olds must be mathematical in nature.

Another example involves table tennis. Students are told:

The ball has an average speed of 11 m/s
The ball takes 0.25 s to travel the same distance as the length of the table.
Calculate the length of the table.
Use the equation: distance travelled = speed × time

They are told that the answer is in m. It is not clear why they would want to do this calculation, nor who would try to calculate the length of the table from the average speed and the time, when it would be far more accurate – not to say straightforward – simply to measure the length with a metre ruler or measuring tape. Again, students are not asked to do anything with the results of this calculation; it is just putting numbers into a formula and turning the handle for the sake of it. It does not assess whether students can think mathematically; nor does it encourage teachers to teach their students to think mathematically.

This is not an isolated question. Another example is:

One of the cells in Figure 2 is 12 mm in length in the microscope image.
The size of the real cell is 0.03 mm.
Calculate the magnification of the microscope.
Use the equation:

$$\text{magnification} = \frac{\text{size of image}}{\text{size of real object}}$$

Once again, students are not expected to do anything with this information. Nor is there any discussion of the fact that in order to know the size of the real cell, the magnification of the microscope must be known. It is, and can only be, just a calculation for the sake of getting students to do some mathematics; never mind that it makes limited sense scientifically.

There are more challenging questions in the higher tier paper; for example:

Sulfur dioxide is an atmospheric pollutant.
Sulfur dioxide pollution is reduced by reacting calcium oxide with sulfur dioxide to produce calcium sulfite.

$$CaO + SO_2 \rightarrow CaSO_3$$

7.00 g of calcium oxide reacts with an excess of sulfur dioxide.
Relative atomic masses (A_r): O = 16   S = 32   Ca = 40
Calculate the mass of calcium sulfite produced.

Students must calculate the relative formula masses of calcium oxide and calcium sulfite and either calculate number of moles or use ratios to come up with an answer. However, a significant proportion of students in this age group would not be able to begin to tackle this question, which is why questions of this nature are only found on the higher tier paper.

What do teachers make of this, and what should their response be to the increases in mathematics content within the science papers? There is again limited research evidence to draw on in answering this question. However, the number of questions which follow the pattern *some information, an equation, substitute to find the answer* surely encourages teaching students to do just that. In the school where I teach, the majority of lessons encourage students to think mathematically, but students are given a lesson prior to their external assessments where they are encouraged to highlight the numbers in a question, write the numbers over the equation, rewrite the equation and put the numbers into the calculators as written in the equation. The booklet they are given instructs them to ignore the context to the question and encourages them that even if they do not understand that context, they can still plug in the numbers and answer the question.

To be clear: this is not to criticise teachers. They are showing that they understand the assessments students will sit, understand how daunting students will find them and are giving students the confidence and skills to tackle questions they find challenging. They are supporting their students to get the best grades that they can. I have taught this lesson myself and it was very popular, with students' confidence improving as a result. But it is not teaching students to think or reason mathematically.

Perhaps the curriculum is simply too demanding for a significant number of students, and more research is required to understand whether and how students can indeed to be taught to think mathematically – and how this can be appropriately assessed.

The amount of mathematics within science

There were four drivers which led to an increase in the amount of mathematics within science from 2015: concern about standards; transition to further and higher education; that school science should be authentic; and to improve students' mathematical confidence and achievement (Wong, 2019). Inherent in the last driver is the notion of transfer, that students will transfer their confidence and understanding from the science to the mathematics classroom. There is very limited research evidence to support the notion that more mathematics within science leads to an increase in students' achievement in mathematics. This is not to say that it does not do so, but the idea has not achieved much research scrutiny. Most research about transfer between mathematics and science has focused on transfer from mathematics to science.

What amount of mathematics 'should' be present within science? And what mathematics should be included in the science curriculum? Does the increase to 20% of the curriculum lead to students learning to think mathematically within science? And if so, how can teachers best support the development of such thinking – in all students, not just those who find mathematics and thinking mathematically straightforward? What actually happens in science classrooms? Does the 20% of the curriculum which is now mathematics within science represent a good use of classroom time? Is it worthwhile science? What other valuable learning opportunities or scientific practices are being sidelined in the quest for more mathematics within science? Does it actually lead to a more rigorous science curriculum with higher standards, whatever that actually means? Is it a curriculum which is rewarding for all students to follow, given the study of science is compulsory up to the age of 16?

Conclusions

There are no easy answers to the questions raised in this chapter, which is why the position of mathematics within the science curriculum remains contested and a matter for debate. There are few answers from the research literature for teachers asking how to support students to think mathematically within science. There is a need for further research to understand how students bring mathematics to bear on problems in science, how they can be supported to think mathematically and the types of activities in which teachers can most profitably engage within the science classroom. If the increase in mathematics within science is justified on the grounds that it will support students' learning within the mathematics classroom, then there is also a need for a research agenda which explores whether this is indeed the case.

Further, there is a need for curriculum designers to ensure that any curriculum which has to be studied by all students is accessible to all students. The types of activities likely to be encouraged by the curriculum and national assessments should be worthwhile and allow all to feel a sense of progress and achievement. Science education for all students can only be justified when it offers something of value to all students, not just those who will get the highest grades.

Questions for further debate

1   What mathematics is it appropriate to expect of students at different stages within science education?
2   How can students be supported to 'think mathematically' within science and what learning activities are the most appropriate to support this learning?

Suggested further reading

Mestre, J. P. (Ed.) (2005). *Transfer from a modern multidisciplinary perspective*. Information Age Publishing.
Redish, E. F., & Kuo, E. (2015). Language of Physics, Language of Math: Disciplinary Culture and Dynamic Epistemology. *Science and education, 24*, 561–590.
Wong, V., & Dillon, J. (2019). 'Voodoo maths', asymmetric dependency and maths blame: why collaboration between school science and mathematics teachers is so rare. *International Journal of Science Education, 41*(6), 782–802.

Note

1   All school science questions are taken from AQA combined science papers via Exampro.

References

Boaler, J. (2002). The development of disciplinary relationships: Knowledge, practice and identities in mathematics classrooms. *For the Learning of Mathematics, 22*, 42–47.

Dodd, H., & Bone, T. (1995). To what extent does the national curriculum for mathematics serve the needs of science? *Teaching Mathematics and Its Applications*, 102–106.

Goldsworthy, A., Watson, R., & Wood-Robinson, V. (1999). *Getting to grips with graphs*. ASE.

Honey, M., Pearson, G., & Schweingruber, H. (2014). *STEM integration in K-12 education: Status, prospects, and an agenda for research*. National Academies Press. www.nap.edu/catalog.php?record_id=18612

Koenig, J. (2011). *A survey of the mathematics landscape within bioscience undergraduate and postgraduate UK higher education*. UK Centre for Bioscience, Higher Education Academy. www.bioscience.heacademy.ac.uk/ftp/reports/biomaths_landscape.pdf

Larsen-Freeman, D. (2013). Transfer of learning transformed. *Language Learning*, 63(Suppl. 1), 107–129.

Lave, J., & Wenger, E. (1991). *Situated learning*. Cambridge University Press.

Leinhardt, G., Zaslavsky, O., & Stein, M. (1990). Functions, graphs and graphing: Tasks, learning and teaching. *Review of Educational Research*, 60(1), 1–64.

Lerman, S. (1999). Culturally situated knowledge and the problem of transfer in learning maths. In L. Burton (Ed.), *Learning mathematics: From hierarchies to networks* (pp. 93–107). Falmer.

Lobato, J. (2012). The actor-oriented transfer perspective and its contributions to educational research and practice. *Educational Psychologist*, 47(3), 232–247.

Needham, R. (2016). Mathematics in science. *School Science Review*, 97(360), 14.

Ofqual. (2015). *GCSE Subject Level conditions and requirements for science*. Ofqual. https://assets.publishing.service.gov.uk/government/uploads/system/uploads/attachment_data/file/819655/gcse-subject-level-conditions-and-requirements-for-combined-science.pdf

Osborne, J. (2014). Teaching scientific practices: Meeting the challenge of change. *Journal of Science Teacher Education*, 25, 177–196.

Porkess, R. (2013). *A world full of data – Statistics opportunities across A-level subjects*. Royal Statistical Society and The Institute and Faculty of Actuaries. www.rss.org.uk/site/cms/newsarticle.asp?chapter=15&nid=125

Redish, E. F. (2017). Analysing the competency of mathematical modelling in physics. In T. Greczylo & E. Debowska (Eds.), *Key competences in physics teaching and learning* (pp. 25–40). Springer.

Redish, E. F., & Kuo, E. (2015). Language of physics, language of math: Disciplinary culture and dynamic epistemology. *Science and Education*, 24, 561–590.

Schwartz, D., Bransford, J., & Sears, D. (2005). Efficiency and innovation in transfer. In J. P. Mestre (Ed.), *Transfer from a modern multidisciplinary perspective* (pp. 1–51). Information Age Publishing.

Turşucu, S., Spandaw, J., Flipse, S., & de Vries, M. J. (2017). Teachers' beliefs about improving transfer of algebraic skills from mathematics into physics in senior pre-university education. *International Journal of Science Education*, 39(5), 587–604.

Venville, G., Wallace, J., Rennie, L., & Malone, J. (2002). Curriculum integration: Eroding the high ground of science as a school subject. *Studies in Science Education*, 37(1), 43–83.

Watters, D., & Watters, J. (2006). Student understanding of pH. *Biochemistry and Molecular Biology Education*, 34, 278–284.

Wong, V. (2018). *The relationship between school science and mathematics education.* King's College London.

Wong, V. (2019). Authenticity, transition and mathematical competence: An exploration of the values and ideology underpinning an increase in the amount of mathematics in the science curriculum in England. *International Journal of Science Education, 41*(13), 1805–1826. https://doi.org/10.1080/09500693.2019.164 1249

Wong, V., & Dillon, J. (2019). 'Voodoo maths', asymmetric dependency and maths blame: Why collaboration between school science and mathematics teachers is so rare. *International Journal of Science Education, 41*(6), 782–802.

Index